No. 990
$7.95

THE
DREAM HOUSE THINK BOOK
BY S. BLACKWELL DUNCAN

TAB BOOKS
Blue Ridge Summit, Pa. 17214

THE
DREAM HOUSE
THINK BOOK

FIRST EDITION

FIRST PRINTING—OCTOBER 1977

Copyright © 1977 by TAB BOOKS

Printed in the United States
of America

Library of Congress Cataloging in Publication Data

Duncan, S   Blackwell.
   The dream house think book.

   "Tab   990."
   Includes index.
   1. House construction--Amateurs' manuals.
I. Title.
TH4815.D86   690'.8   77-21197
ISBN 0-8306-7990-1
ISBN 0-8306-6990-6 pbk.

Foreword

Today, far too many houses look as though they were all punched out by the same die, characterized by a dull box-and-lid monotony, as though families were all alike. This book is for people who want to plan a house that *isn't* like all the rest, a house that meets specific family needs without costing a mint. It's a step-by-step guide to generating your own design ideas, to drawing your own plans, to working out the details of every wall and floor of your dream.

There's a whole chapter on choosing a homesite. It's a tough-minded course in how to get the piece of earth you want—for a price you'll be willing to pay. It takes a close look at building codes, zoning restrictions, waste disposal, percolation tests, water sources, land investment, taxes, you name it.

But the real meat of the book is about planning your dream house. It teaches you how to start with design ideas—all your own—and how to sift them out, refine them, mold them into real plans that work on paper and on your site. There's plenty of experience-based information on architectural styles, types of construction, floor plans, house siting, plot plans, interior design, and lots more.

Contents

Choosing a Homesite

1

Just about anyone can design his own home. And this means not just a house—heaven knows there are enough house plans floating around to fill a 10-ton truck—but a home, in the full sense of the word. We're talking about your castle, your dream home. Call it what you will, it's the place where you and yours will spend the better part of your lives. You might just as well have something that you like, can enjoy, and take pride in—and will fulfill your needs. All of them. Why settle for second best, especially when you don't need to? But to build that home, you've got to find someplace to put it.

URBAN, SUBURBAN, EXURBAN, AND RURAL

Ideally, the design of a home should begin not with floor plans or architectural "styles" but with the ground upon which the home will rest. There are two approaches here. One: the prospective homeowner can search for a suitable piece of property upon which he would like to build a house someday. Two: the prospective homeowner can first determine the sort of home he would like and then scurry about locating some property on which to build it. Either way, the site is, or certainly should be, an integral part of the design of the home. The location and the immediate environs should complement the structures built thereon, and conversely, the structures should complement the surroundings and not stick out of the

9

landscape like some sort of spurious growth. Both elements should fit nicely together in harmony. And as you will see, this holds not just for appearances, but for other quite practical matters as well.

To take this a step further, there is a strong three-way relationship involved here. In any completely successful home design the three foremost factors are: general location and immediate surroundings, design of the structure, and life-style of the inhabitants. The right house for the right people in the wrong location becomes less than fully successful. The right house in the right spot inhabited by the wrong folks becomes untenable. A true home defies the laws of mathematics because the whole is greater than the sum of the parts. Complete compatibility of all the parts is what makes this possible.

For the sake of simplicity, then, let's assume you want to design and build a home but you have no property. You have nothing beyond a few vague ideas about structural design, and you have no concept of how to go about choosing an appropriate site.

The *perfect* site for your home, or any home, simply doesn't exist. However, you have to start somewhere. Choosing your homesite will involve some thought, and then some compromise. Perhaps a lot of compromise. But to begin with, you must settle upon a general location, chosen from some point in the world. Unless you are contemplating a truly major change, you will probably unhesitatingly narrow your choice to somewhere in the United States. This then will be narrowed, most likely but not necessarily, to your own region—New England, Pacific Coast, Midwest, or wherever. You may opt to remain in your own state, or you may not. Most of these decisions will be influenced by factors far outside the realm of house design.

Then comes a further narrowing down. When you start to look for your own patch of ground, you have four basic categories from which to choose: urban, suburban, exurban, and rural. None but totally urban or totally rural areas are strictly delineated.

Each area does have certain characterisitcs, however. Urban means, to most folks, in the city. Urban areas have a substantial population base with a high density, a good deal of commerce, concentrated social and cultural amenities, an assortment of business establishments. Building sites are generally rather small, sold by the lot or the square foot. House building is restricted and regulated. On the other hand, municipal services such as fire and police protection, rubbish removal, and the like, are at the maximum. Proximity to one's work, cultural opportunities, social affairs, and the possibilities for civic involvement are a part of the urban scene.

Suburbia, much maligned and much praised over the past few years, consists of sizeable residential areas outside the main urban sprawl. There is little commerce by comparison with the city. Building sites tend to be somewhat larger, sold by the lot at somewhat less cost. The taxes are usually a bit less strenuous, and more flexibility is allowed in building, although there is usually considerable regulation and restriction here as well. Municipal services are more limited, and commuting becomes a way of life.

Exurban areas are just beyond the suburbs, not close enough to the city to be in the suburbs but not really out in the country either. Property here comes in large parcels. The cost of land, especially if reckoned on a per-square-foot basis, is considerably less than urban ground, and taxes are also usually less. There is far more room to swing one's elbows, building siting is flexible, and the building regulations are likely to be fewer. Services are at a minimum, and the exurbanite must provide his own transportation.

Rural areas are farthest from the cities. Urban contact is almost nil except at individual whim. Land is sold in parcels of about any size, even several thousand acres. Practically speaking, there are usually no services which can be depended upon fully, including snow removal, road grading, medical assistance, and fire protection. When you live in a truly rural area, you might as well figure that virtually everything in life, even entertainment, is pretty much up to you and your family. Even if the school bus *is* fairly regular.

Now you have to decide which of these areas would serve your purposes and your life style the best, and that decision may not be easy. There are so many considerations to think about, and only you can sort them all out. You'll have to live with your decision for a long time. By comparing, balancing factors, trading benefits for deficits, and swapping more important points or characteristics for those which rank lower on your scale of values, you will eventually come to a decision as to which area is best for you.

NARROWING DOWN THE POSSIBILITIES

How do you go about narrowing down the possibilities? How do you figure out which area is best for you? It's not an easy job. All I can do is get you started. It's a process of building up, sorting out, and boiling down. Oftentimes it is easier to determine what you *don't* like, rather than what you do. And that's as good a way as any to get the wheels rolling. Stick your feet up on the coffee table, grab a sheet of paper, and start a list of things that you simply *won't* condone. Here's a sample list:

NO WAY

1. Visible high-tension lines and towers
2. Visible telephone poles and lines
3. Big buildings within sight
4. Freeway noises
5. Busy highways in the immediate vicinity
6. Too far from work

And so forth. Add as many negatives as you can come up with.

When you run out of steam on those, start up another list of things that you would like to have. Let your mind soar in glorious abandon, and for the time being at least, forget about economics and assorted practicalities. Like this:

SUPER IDEAS

1. Plenty of privacy
2. Spectacular view of mountains
3. Trout stream within 10-minute walk
4. Private supply of sweet water
5. Room to raise and graze two cows
6. Plenty of tall trees
7. Sun from early morning to late afternoon
8. A swimming hole
9. Plenty of woods
10. Located at end of gravel road

And so on. Just let the ideas flow.

When you complete all this seeming nonsense, go back and analyze what you've put down. For instance, what does "No big buildings within sight" mean? Is that a 40-story skyscraper 4 miles away? Or a four-floor hotel four blocks away? Or is it your neighbor's large house just down the road? And then there's "Too far from work." Is that two minutes, or two miles?

And then there's the privacy angle, too. What is that? Plenty of tall trees could be a forest or a half dozen old elms. The amount of ground you need for two cows depends upon a number of factors. You'll have to define exactly what your preferences imply.

Okay, let's say you've decided on the geographical area where you want to settle. And you've made up a list of preferences. Before you can decide on one of the four categories within that geographical area, you'll have to gather some facts. For instance, how far would you have to drive to work from different areas? How about shopping areas? How far away are they? What about the future? Get out your crystal ball and do a little gazing. You might even want to get some facts and figures from the Chamber of Commerce and City Hall to help you along with this part of the project. What's happening in the areas you're considering? Will there be industrial expansion? Housing developments? Are property values going up or down? Try to judge what the environs will be like 5, 10, even 20 years from now. And of course you'll have to determine whether or not you plan to be there X number of years into the future. Maybe you want to situate so that you can sell 10 years hence.

So after gathering some facts and sifting through your preferences, you should be able to determine which category is for you.

SOME THINGS TO CONSIDER

All right, let's jump ahead another notch. Let's assume that you have finally come up with two or three or a half dozen parcels which seem to fit your preliminary requirements. Now you have to choose one of them. As the number of plots narrows down, the questions you ask yourself should become more and more pointed and detailed. The object is to get everything you can, give up as little as possible.

There are certain basic questions which can be asked about nearly any piece of ground. Perhaps the most important question is, does the site excite you? Does it do anything for you? It should. A site or a parcel must have something that arouses your interest, excites your imagination, perks up your thoughts. If the whole tract just lies before you like a lump and doesn't stir up any feelings now, it probably never will. You'll become less and less enchanted with the spot as the months go by.

How about siting possibilities? On a small city lot there may be only one place to put a house. A large lot may offer two or three spots. On a 50-acre parcel there could be 50 or 100 spots.

Some parcels will have one or two good sites for buildings, along with several other marginal spots which could be used with a little extra work. In other cases a considerable amount of work must be done in order to gain a decent site at all. And in a few cases, the site is already prescribed. Some governing body, such as a subdivision homeowners' association, will have exercised their collective wisdom, done a bit of surveying, and driven a stake at some particular point in each lot. The would-be homeowner must then construct his house centered no more than 50 or 75 or 100 feet radially from that stake.

The more siting possibilities there are, the more flexibility you will have in establishing your home just the way you want it on the property. The fewer there are, the more restricted you will be. Of course, if there is just one, and you think that one is simply terrific, then you're all set.

You also have to consider noise. Traffic noises, children playing in the schoolyard at recess time, jets thundering over from a nearby airport, that sort of thing. Some people can't get along without the sound of rushing traffic. I was once visited at my farm by a friend who wanted to get out of the city and relax for a while. He loved the place—and left two days later. The silence in the daytime was driving him wild, and he couldn't sleep at night for the crickets chirping and the frogs croaking. Other people must have absolute solitude. So what you must do is determine just what quiet means to you.

There are a number of specifics you can look at. Where are the highways in relation to the parcels? Traffic noise can carry long distances from a freeway. If the noise is steady and constant at some particular level, you can grow used to it. If it is spasmodic and changeable, it may be distracting. Remember, noise levels in the summer (with leaves on the trees) are quite different from those in the fall and winter. The prevailing winds, which in many places reverse themselves at dawn and dusk, may also influence noise levels.

Check the surroundings. How close are the neighbors? (How close is close?) Can you see them? Do you want to see them? Can you hear them? Do you want to hear them? Will there be noise from a nearby municipal swimming pool, tennis courts, drive-in movie, trucking terminal, schoolyard? Investigate everything around the parcel that might have some impact upon your life. Try to look ahead, too, and determine what might happen later. And don't forget the airport; you could end up right under a landing or holding pattern.

A prime consideration is security. If you build a house on this lot, or that parcel, will you feel secure there? That feeling of security includes a lot of factors, many of them subconscious. Either you feel secure or you don't, and no amount or rationalizing will vaporize a feeling of insecurity.

You've got to ask some more questions. Are there enough people around to make you feel comfortable? Or are there too many, so that you feel bothered, that they are watching you. Are there neighbors or others close enough at hand so that you can quickly reach them in an emergency? Or are you self-sufficient enough not to be concerned about that? How about fire protection? If you live in a city, fire protection will be practically guaranteed. But as you move away from the city, protection becomes more and more problematic. People who live in the country must be more conscious of fire safety. They must be equipped with fire extinguishers and should have a fire and smoke detection system installed in the house.

Then there's police protection. If police protection is very important to you (and it is to most people), you will want to consider this factor before buying a plot.

The proximity to medical assistance is another consideration. Your parcel of real estate may be 5 minutes from the nearest hospital or doctor, or 5 miles, or 5 hours. Again, you will have to find some compromise that you can live with comfortably and without worry.

You also must consider day-to-day convenience. How far away from stores, shops, gas stations, etc. do you want to live? It's a matter of dependency, or independency, of reliance upon yourself, or upon others. How much of either can you live with, be comfortable with, afford to have? Be honest with yourself. If you want to be catered to, the boonies will bust you. If you prefer to take care of all your own needs, you may be miserable in the city.

You may also want to consider "the view." Everybody who wants to buy a piece of ground wants a view to go with it,

Fig. 1-1. A "view" is in the eye of the beholder. Here, the foreground is only a few feet away, the midrange about a mile, and the long range extends to about 20 miles.

16

and of course it always must be a "good view." But there are two different ways of looking at views. Like a photograph or a painting, a view may have foreground, middleground, and background. Or a view may consist of just one of these elements. Most people feel that a proper view should have all three elements and should extend over some distance, rather than being compressed into just a few yards. (Such a view is shown in Fig. 1-1.) But others prefer a view which gives them a closed-in feeling, a view that has only one element.

But whatever kind of view you want, remember that it is only one factor. In choosing a plot, you must weigh *all* the factors.

And another one of those factors is access. In order to enjoy your homestead, you have to be able to get to it easily. There should be at least one place, and preferably more, where you can build a driveway or road to the house or garage site. If you are desperate for a horseshoe type drive, you will need two suitable access points, one to drive in and the other to drive out. In some areas, notably cities and subdivisions, each plot may be allowed only one curb cut—the place where the curbing is cut out to allow the driveway to merge smoothly with the roadway. This means no horseshoe drive, no back entrance or cross drive.

Most people prefer that the drive be short and straight, with a slight slope. This means quick and easy snow removal, good drainage, and low costs for construction. The longer and curvier the drive is, the higher the costs of construction and upkeep.

But access involves more than just driveways. It includes unobstructed roads that lead to your property. Check into the efficiency of the snow removal crews. You might be on a half-forgotten dirt road that becomes snowbound periodically. Inquire about road maintenance, too, to be sure that the road will remain in good shape and won't dissolve during spring thaws. Actually, road surface and snow removal conditions are often better in the country than in the city, but you'd best get a reading on the local situation and plug the information into your list of facts and possibilities.

There is also the question of legal access. If you are looking at outlying land, make certain that the road leading to that land is a recorded, deeded right-of-way, legally open to public use at any time. Don't take anybody's word for it. Look up the recorded documents that prove the fact. There are parcels sold every day to unsuspecting buyers who suddenly discover that their only means of ingress and egress actually belongs to somebody else.

Many parcels, often lovely ground with small price tags, are completely landlocked. They are surrounded on all sides by privately owned land. In such a case you have two options. You can buy an additional strip of land, wide enough for a road, between your landlocked parcel and the nearest public road. Or, you can get an easement or right-of-way from the intervening landowner(s) for the same purpose. Either way, you will have to build a road. And either way this can be time consuming, frustrating, and expensive. If you decide to go ahead, work out the details with a competent real estate agent and a good lawyer. Get everything down in black and white, in legalese, and record the documents.

While we're talking about legalities, let's look at zoning. Before buying a parcel, find out how it is zoned and then determine exactly what the regulations are and how they will affect your building, your contemplated designs, your activities, and the value and the resale of the property. If there is no zoning, see what the likelihood is for future zoning.

The lack of zoning, or weak zoning, has engendered something known as protective covenants. Some are good and some are bad, and sometimes there is a question of just who is being protected. Covenants are usually found in subdivisions and in some housing developments or tracts. They are usually drawn up by the developers themselves, and sometimes by the homeowners. Properly established, convenants become a part of the legal documents which change hands when the property is conveyed, both originally and forevermore. And whether or not you are aware of them, once you sign on the dotted line and become an owner of property protected by covenants, you must abide by them.

Most of the clauses of covenants are general and not terribly restrictive, though the opposite can just as easily be true. You may see restrictions like these: "No above-ground fuel tanks; all clotheslines must be surrounded by a 5-foot fences; horses or other domestic stock not allowed; single-family residences only; buildings must be placed 20 feet from lot lines; garages should be no larger than two-car size." So it's best to check what's on record.

Architectural controls, usually administered by a committee of homeowners and/or developers, should also be looked into before buying land. Usually such restrictions are found in exurban high-rent districts, though they can be put into effect in any controlled area. These restrictions regulate certain architectural practices. They may prohibit the use of certain building materials, prescribe lot clearances, dictate the size of dwellings, etc.

And, of course, you should check out the building codes. You may find a couple of different ones. Go over the details of the codes carefully and see if you can live with them. Then, once you commit yourself, follow them faithfully and cheerfully. You'll get lots more cooperation from the authorities and have far fewer frustrations.

Then there is the problem of liquid waste. In urban areas, and in most suburbs as well, there is a sanitation department to take care of such matters. The homeowner simply plugs into an existing sewage line, and that's all there is to it. The cost for this service can be fairly high in comparison with other methods, though you don't notice the bite as much. Usually there is some sort of tap fee involved. This may be as low as $200 but can easily run as high as $2000 or more. The fee may be included in the purchase price of the real estate as a hidden cost or may be specifically spelled out or may be separately payable at tap time. And that could be a shocker if you didn't know the details beforehand. And then from the time you tap onto the system until you sell the property, some small portion of your municipal taxes will be diverted to the operational expenses of the entire system.

In some areas, you will find small waste disposal systems which are operated in roughly the same manner as larger municipal systems but are actually owned by a developer, a homeowners association, or a private company. Tap fees are usually assessed at the time of purchase of the real estate. But instead of the system being supported by means of your tax dollars, a periodic direct assessment is made to each homeowner. Oftentimes the authority to make these assessments is open-ended, which means that the homeowner can get royally socked for some substantial sums. Check this one out thoroughly, too. Get a reading on costs and what sort of protection or guarantees the homeowner has after he ties into the system.

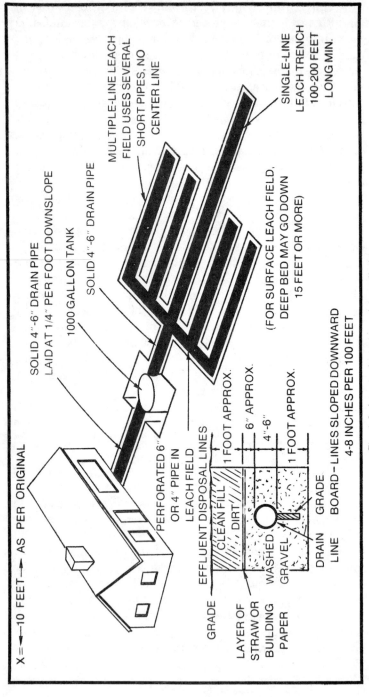

Fig. 1-2. A typical septic tank installation.

SOLID 4"–6" DRAIN PIPE LAID AT 1/4" PER FOOT DOWNSLOPE

1000 GALLON TANK

SOLID 4"–6" DRAIN PIPE

MULTIPLE-LINE LEACH FIELD USES SEVERAL SHORT PIPES, NO CENTER LINE

SINGLE-LINE LEACH TRENCH 100–200 FEET LONG MIN.

X = 10 FEET — AS PER ORIGINAL

PERFORATED 6" OR 4" PIPE IN LEACH FIELD

EFFLUENT DISPOSAL LINES

(FOR SURFACE LEACH FIELD, DEEP BED MAY GO DOWN 15 FEET OR MORE)

GRADE

1 FOOT APPROX.

6" APPROX.

4"–6"

1 FOOT APPROX.

CLEAN FILL DIRT

LAYER OF STRAW OR BUILDING PAPER

WASHED GRAVEL

GRADE BOARD — LINES SLOPED DOWNWARD 4–8 INCHES PER 100 FEET

DRAIN LINE

20

Semirural and rural residents have to take a different approach altogether. Today, almost everywhere, regulation by state health departments is rigorous, and what you can and cannot do in the way of waste disposal is rather clearly spelled out. You have two principal alternatives: a septic tank or a domestic sewage disposal system.

There is a considerable difference in how the two systems work, though the end result is about the same. A septic tank is simply a large holding tank, either of steel or concrete, which receives the liquid waste directly from the house waste lines (Fig. 1-2). Bacterial action in the tank breaks down the solids. Mineral particles fall to the bottom of the tank to form a sludge, which must be pumped out every so often. The gasses that are formed escape through tank vent pipes, and the rest remains liquid (Fig. 1-3). Whenever raw sewage is admitted to the tank, an equal volume of liquid effluent flows out of the tank outlet pipe, so the level in the tank remains constant. The effluent, rendered relatively harmless by the bacterial action, runs into a leach field, a series of perforated pipes laid in gravel. Most of the liquid sinks into the soil, some evaporates from the top of the field, but all eventually ends up in a water table somewhere in a purified state.

The domestic sewage system also employs a main holding tank but includes some mechanical apparatus. Sewage is

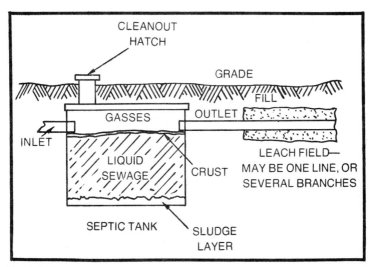

Fig. 1-3. Cutaway view of a sewage disposal system using a septic tank and leach field.

broken up, aerated, passed out into a leach field, and recycled. The resulting effluent, however, is purer than that from a septic tank, and there are other claims made for longer life, less trouble, and lower cost.

All this brings up a few points for the prospective land buyer. First, if there is no municipal system available to tap onto, then obviously you must provide your own system. This means expense. Somebody will have to dig a hole for the holding tank and trenches for a leach field. You will need a tank, a suitable amount of pipe, and several truckloads of gravel. Once the equipment is in place, somebody will have to backfill around it. You will be able to save some money by doing some of the work yourself, or you can get an estimate from a contractor (or better yet, two or three contractors) and have him do the whole job.

Second, you will need space on the lot where the system can be placed. The main tank is usually placed within 10 feet of the house (as close as possible to the most centralized plumbing location within the house). The leach field then must start about 10 feet from the tank. The tank hole required will probably be about 10 feet square. Setting the leach field will mean tearing up a large piece of ground, the exact size being dependent upon the characteristics of the soil (Fig. 1-4). But you can probably figure on needing an area about 10 feet wide and 100 to 200 feet long for the field.

Third, the location of that necessary space is important. You've got to be able to dig a hole deep enough for the main tank, which may be 5 or 6 feet high and about 18 inches below grade. And you've got to be able to dig the leach field, too. All this means that if there is bedrock or huge boulders close to the surface, you are going to run into a lot of problems in getting the system into the ground. Digging rocky ground is expensive; blasting, even more so. If the parcel is small, you may not be able to get the house and the septic system properly situated. To compound the problem, the leach field must slope gently downhill away from the house, away from any outdoor living areas like pools or patios, and away from drives or walks. In addition, any surface leach field must be positioned so that no vehicles will drive over it and crush the pipes, and it must be well separated from any water supply like a well or spring. It must be away from streams, low spots, marshy ground, or runoff water courses where the ground

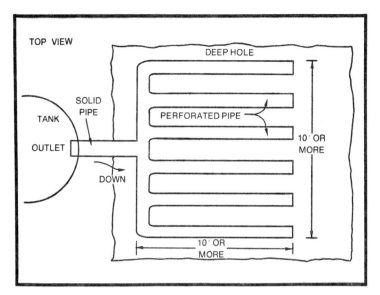

Fig. 1-4. A typical leach field network.

may become saturated in wet weather. All parts of the system should be accessible to heavy machinery so that the tank can be periodically pumped out and the leach field dug up and replaced when necessary. And on some pieces of ground, putting all these requirements together can be nearly impossible.

Fourth, the effectiveness of a septic system depends entirely upon the soil's ability to absorb water. This ability is called *percolation*. As the effluent runs out into the leach field, it must leave the pipes and percolate downward through the underlying strata. The rate at which it does so is measured by making what is referred to as a percolation test. In most areas you must obtain a permit to install a septic system. The first step in obtaining a permit is a percolation test. Only after the percolation rate has been shown to be satisfactory will the permit be issued. So if you buy land that can't pass a percolation test, you may not be allowed to install a septic system.

I knew a man who bought a lovely 2-acre parcel of rural land which had everything he wanted in the way of a homesite. He paid cash and started to build his dream house. Later, he found out that his beautiful land couldn't pass a percolation test. No septic system. No dream house. No resale value.

So, make those tests *before* you buy, not after. Or ascertain that the tests have been made and that the results were positive and that they are on file at the local health department or building department offices. Or make sure that a positive percolation test is a written part of any sales agreement or option or purchase contract that you sign. No perc test, no buy.

You may even want to run your own percolation test. It's a fairly simple job, but make sure the proper official is on hand to verify the test results. Requirements and standards vary from place to place, but in general the routine goes something like this. First spot the most likely area where the leach field will be built. Dig a hole about 3 feet across to give you some working room; dig it to the same depth that the leach pipes will be (Fig. 1-5). Follow the local requirements for this—the depth will most likely be about 2 feet. Make the bottom of the hole flat and level. Then dig another hole in the center of the first one, 1 foot square and 1 foot deep. The next hassle is the water. You'll need a lot, and you may have to freight it in by the trash can full. Depending upon local requirements, you may have to keep that cubic-foot hole topped with water for 12 or 24 hours, and periodically check the rate at which the water disappears. If the water level lowers at a rate of about an inch

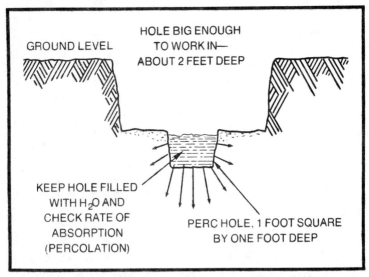

Fig. 1-5. A typical percolation test arrangement. Specific details and requirements depend upon local regulation.

per minute, the percolation is excellent. An inch in 5 minutes is not too shabby, and an inch in 10 minutes may be acceptable. If the rate is an inch in an hour or longer, you may be completely out of luck.

If you're going to buy some land, you should also consider the water supply. If your parcel is in a city or a suburban area, you won't have to worry much about where the water is coming from. But in some exurban areas there is a municipal water supply and purification plant, with a water department to run the system. Costs for obtaining water can be high, and the quality of water low. The cost involved usually includes a tap fee to hook onto the system, followed by regular monthly or quarterly billings, either flat-rated or based upon meter readings.

Some water systems in exurban and rural areas come with mixed blessings. These systems are usually smaller and likely to be less efficient than their city brethren. Breakdowns may be more common; water pressure may fluctuate. On the other hand, the water quality is likely to be better. And the costs, which usually consist of tap fee plus direct periodic billings, are usually lower. Some of these systems produce soft water; most produce very hard water. And if you hate hard water, you may have to buy a water softening system.

If there is no water main near your property, then you will have to provide your own water system. It might be expensive initially, but there are advantages. You will have complete control over your own system (except for whatever regulations may govern it), and over the long haul the chances are excellent that you will pay less for a higher quality of water than with any other type of system.

There are several ways to obtain water on your property (Fig. 1-6). Some methods are a bit more of a gamble than others. And you just never know how long the source will last.

In all cases , however, the starting point lies with the law. There may be water on or under your real estate, and you may not own a bit of it. In many areas, particularly in the West, water rights are strict and are rigorously enforced. Water, for whatever the purpose, cannot always be taken from a handy stream by just anybody because somebody else may have prior claim to it. The same goes for still bodies, such as ponds and lakes. And water from irrigation ditches, which in most cases only run part of the year anyway, can only be used by those who own the water rights, or by special permission.

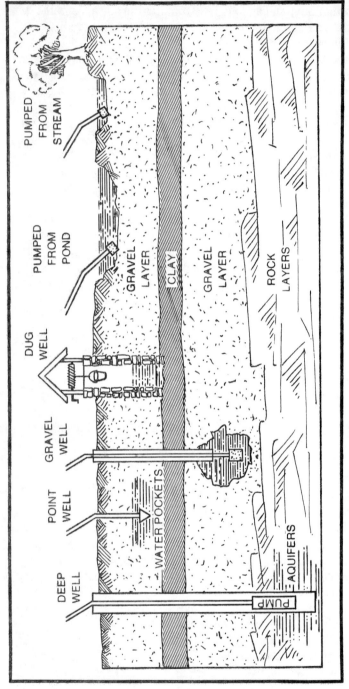

Fig. 1-6. Six principal sources of domestic water for those who cannot tap onto a municipal supply.

26

Water from a permanent spring or a fresh pond may have been conveyed or excepted by deed by the seller during a property conveyance. Much more common are state requirements which specify that a property owner must have a permit before he can dig any kind of well.

The first step, then, is to find out what water rights you will or will not have when you purchase a piece of property. Find out if a permit is necessary, and if so, whether or not you will be able to get one. Then, if you're really serious about buying the parcel, you can investigate water supplies and methods of using them.

Investigate the *visible* sources. A good spring makes a marvelous source, provided you don't put too heavy a demand upon it. You can do some asking around to find out how constant the flow is. Check it yourself to determine the approximate regular flow

A fresh pond or a constant stream is also frequently used for a water source. A strainer-foot valve positioned below low-water mark, a length of pipe, and a suitable pumping arrangement will put the water into your home. Usually a certain amount of filtration and purification equipment is also necessary. Standing surface water, as in a swamp or marsh, probably will not be usable but may be worth checking out if nothing else presents itself.

In all cases where surface water is available, you can have the water tested *before* you buy the property. All you have to do is present a gallon of undisturbed water in a sterile container to the local or state health department labs. Or you can get a report from a private lab for only a few dollars. The results will tell you exactly what is in the water, whether the sample passes the state health standards, and what sort of purification and filtration must be done to make the water potable.

The below-ground sources are always something of a gamble. In some places there is a high, fairly constant water table, with good supplies of water found only a few feet below ground in a gravel layer. Volume is sometimes quite good, and water is extracted by means of a point well. A point well consists of a length of pipe with a special strainer-point tip. It is driven into the ground manually until an adequate water supply is reached. Sometimes two or three points, all connected to the same supply line and ultimately to a small pump, are used to increase the capacity of the system.

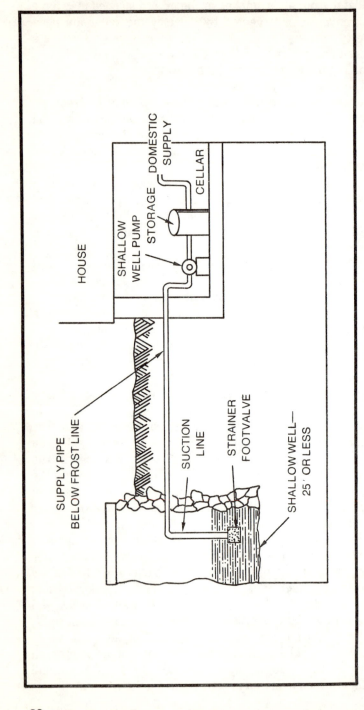

Fig. 1-7. A basic shallow-well water system for residential use.

HOUSE

SHALLOW WELL PUMP

STORAGE

DOMESTIC SUPPLY

CELLAR

SUPPLY PIPE BELOW FROST LINE

SUCTION LINE

STRAINER FOOTVALVE

SHALLOW WELL— 25´ OR LESS

The traditional dug well is just a hole in the ground, from 3 to 6 feet in diameter and 10 to 25 feet deep (Fig. 1-7). The sides are finished with dry-laid stone (i.e., no mortar is used in the joints) to keep them from caving in. And the walls are usually carried a couple of feet above ground for added protection and to keep the cow from falling in. In the old days dug wells had windlasses and oaken buckets, but those times are gone forever. Now dug wells are covered, and the water is pumped up through a pipe that has a strainer-foot valve on the end. A dug well makes a good water supply if you don't work it too hard, but it has an unfortunate tendency to dry up from time to time.

There is another type of shallow well, sometimes called a gravel well. This kind taps a layer of water-bearing sand or gravel close to the surface. A heavy pipe is forced down through the soil until the correct stratum is reached. Then water is forced down the pipe and pumped back up again, forcing out sand and gravel until a cavity is created around the bottom of the pipe. After a period of settling out, the cavity becomes a reservoir of clear water, ready for use. The supply is constant and usually reasonably pure. The cost of establishing and maintaining this type of well is relatively low.

The deep well, or drilled well, is a different type of critter (Fig. 1-8). By means of a rotary drill, which turns constantly and grinds its way down, or a "banger" rig, which thumps into the earth much like a hand-held miner's star drill, a driller drives a deep hole far down into bedrock. The depth may be 50 to over 1000 feet. Once an underground water pool or stream or water-bearing gravel layer is reached and a satisfactory flow rate is established, the well is cased with heavy steel pipe and capped. The necessary supply pipes, strainers, valves, and often the pump itself are installed in the well casing. And as you might expect, all this does not come cheap. The well itself may run $15−$30 a foot. And a pump, not installed, $500 or more. And yet, over the long pull, a deep well may be cheaper to own and operate than a city water tap.

Finding out about below-ground water supplies on a parcel will take some time. Local, county, or state governments may have some information of use to you. Inquire about the water table and how it acts. See what facts they have on file about existing wells in the area. The state geologist can perhaps give you some idea of what might lie beneath a piece of property.

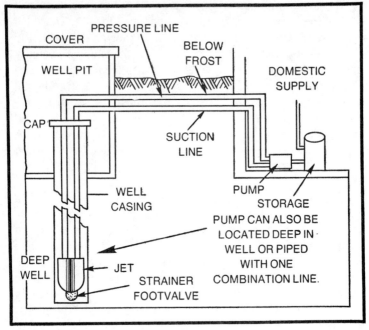

Fig. 1-8. A typical deep-well water system.

Then there are water engineers. You can contract one to do a survey of the property and give you a report of what he finds. There won't be any guarantees, but at least you will have some probabilities and possibilities to work with. And if you can talk the seller of the property into paying for the survey, so much the better.

Here's another thing to consider when you're shopping around for land: telephone service. In urban, suburban, and most exurban areas, service will be reliable and relatively inexpensive. Also, you will probably be able to call over long distances in any direction without incurring any toll charges. In rural areas, though, good telephone service may be rare. The lines may malfunction or go down often. Eight parties may be the smallest group you can get in with; private or business lines may not be available at all. Or if they are, the cost may be prohibitive. You may also be treated to constant ringing of the party line at all hours. If the telephone is an important part of your life, you may decide that a private line and good service is essential. So land in some areas may be out of the question for you.

The proximity of power lines also deserves investigation. The best situation, or the least expensive one at any rate, is when the power lines follow the access road to the property and pass near the proposed site of the house. If the distance between the house and the nearest utility pole is short, then the cheapest way to get power into the house is with a single-span aerial service drop from pole to house (Fig. 1-9). The power company will take care of this at no charge. But for an *underground* service lateral, the closer the house is to the nearest pole the better—because underground service is installed at the expense of the owner. Short runs mean lower wire and installation costs.

If additional utility poles are needed, the cost of each one plus the installation will probably be billed to you, the owner. And in many rural areas, it is not at all unusual to discover a choice piece of land that is miles from the nearest power lines. In this case, you will be charged by the mile for installation of new lines. This is an expensive proposition and may negate a purchase for you.

There are also other services to consider. For instance, is natural gas readily available, or will you have to use oil, propane, or electricity for heating? What about regular delivery of propane, fuel oil, or coal? And then there are such items as library facilities, visiting nurse, parks and playgrounds, mass transportation, dog catcher, and a host of other odds and ends. When you put them all together and decide what you would like to have and what you can do without, you will get a reading on which homesite might be best for you.

Of course, you'll have to think about property taxes too. So how do you figure out what your taxes will be? You don't exactly, until you get the first bill. However, you may be able to get a rough idea by finding out what the taxes are on similar pieces of property in the vicinity, but the picture you get may or may not be accurate. You can visit with the assessor and get a better idea. Tell him which parcels in particular you are considering. Ask him what the taxes might be on those plots with a $20,000 or $40,000 house (plus whatever other improvements you have in mind). His estimate probably won't be right on the dot, but at least you will have a ballpark figure to work with. If the assessor feels that the taxes on property X will be $100 a month and you can't afford that kind of money, then you can scratch that one off your list.

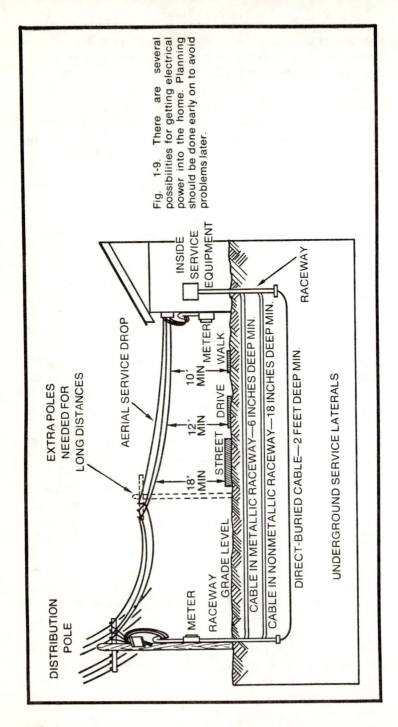

Fig. 1-9. There are several possibilities for getting electrical power into the home. Planning should be done early on to avoid problems later.

DISTRIBUTION POLE

EXTRA POLES NEEDED FOR LONG DISTANCES

AERIAL SERVICE DROP

INSIDE SERVICE EQUIPMENT

METER

18' MIN

12' MIN

10' MIN

METER

WALK

DRIVE

STREET

RACEWAY

RACEWAY

GRADE LEVEL

CABLE IN METALLIC RACEWAY—6 INCHES DEEP MIN.

CABLE IN NONMETALLIC RACEWAY—18 INCHES DEEP MIN.

DIRECT-BURIED CABLE—2 FEET DEEP MIN.

UNDERGROUND SERVICE LATERALS

But even the dollar amount you have to pay is not the whole story. The point is, what are you getting for your tax dollars? Find out where that money goes each year and what you will receive in return for your hard-earned cash.

YOUR INVESTMENT

Buying land is serious business: it involves a lot of money. So when you shell out your hard-earned cash for a patch of earth, you want to make a good investment. You want to buy no more land than you need, and you want it at a fair price. And you want that land to work for you to maximize your investment.

Maximizing your investment? Here are a few examples. If you plan to burn wood in your fireplaces, then a woodlot on your own property could mean big savings. If you want to build a lot of stone walls, then an on-site supply of fieldstone is advantageous. I can testify that having your own gravel bank comes in mighty handy for various purposes. Shoot for every extra advantage and usable feature you can.

Usually the natural features of a lot will have a bearing on where and how you can build, how well you can use the property, and how you can get the most from each dollar you spend. For instance, if you are set upon having a full basement and the proposed site is covered with 6 inches of topsoil over a granite ledge, you will have to blast. The costs will skyrocket. If a considerable amount of tree and brush clearing must be done, as opposed to building upon a clear grassy spot, costs will rise. Large rocks, watercourses, swampy or water-retentive ground, loose and unstable soil, and land which must be recontoured will cost you extra money. The object is to live with whatever is there, or better yet, turn them to your advantage. If you cannot do that, try to avoid the site completely, unless there is no other way to get what you want.

How big should your piece of land be? Obviously, the tract must be big enough for the house. That may sound ridiculous, but many lots aren't. Some towns offer up city lots that measure only 40 feet wide and 80 feet long. Now a house that is 30 feet wide and 60 feet long, offering only a modest 1800 square feet of living space, is going to pretty well use up that lot. And if the local setback ordinance calls for 10 feet from every lot line, you're done before you begin. And even if you *can* build, you'll be able to shake hands with your neighbor from your respective bathroom windows. Watch out for lots that are advertised in terms of square feet, by the way. Go by the actual dimensions.

So you have to have room for the house, the garage, the garden, the pool, the patio, whatever. Plus some breathing space, enough for a lawn maybe, and some more to cover any local setback requirements. These setbacks may be only 5 or 10 feet all around or may call for no structure within 100 feet of the roadway.

Beyond that, what do you need? Nothing really, if you don't want it. But having more land will give you greater flexibility and more options later on when you want to enlarge the house or put in a pool or a vegetable garden or a shuffleboard court.

An acre, for instance, sounds like a whole bunch of ground, but that's not really the case when you move onto it and start spreading around a little. An acre is 160 square rods, which works out to 43,560 square feet. But consider: If that acre were perfectly square, it would measure only 208.7 feet on a side. If you plunk down a 45 × 45 foot house in the center of that acre, you'll have only about 82 feet around each side of the house, and that's not all that much.

Much depends upon the usefulness of the land, how much you want to sprawl out, what you plan to do in addition to building a house, and your ideas of solitude and privacy. And of course there's the cost.

How much should you pay for a homesite? It all depends. Small lots are priced on a basis of so much per square foot. Small tracts, those of an acre or two, are usually priced on the basis of location, desirability, and other nebulous characteristics, at so much per acre. Waterfront land is most often sold on the basis of so much per foot of frontage on the water, with the back land being of correspondingly less cost and value. Larger tracts frequently are priced by the acre, but according to the frontage on roads. Thus, a piece with 200 feet of road frontage and 1000 feet of depth is less expensive than one with 1000 feet of frontage and 200 feet of depth, all other things being equal. In most cases, the price of the property depends upon the social and economic pressures put upon it, the desirability of the property as a place to live. Instances in which the price is dependent upon what crops the land can produce are few and far between these days, except for huge tracts in agricultural or ranching areas. Only you can determine whether or not the price is good or bad, whether the price and the value compare favorably. But remember, that price may well have a bearing, as does the property itself, upon your building program. Consult your innermost being. Then consult your banker.

THE TIME TO SHOP

There is a right time and a wrong time to go looking for real estate. There's an old saying that you should never buy a used car on a dark rainy night. They all look good then: black and shiny and slick and beautiful. But when the sun comes out and the water dries off—woe and gloom. They just don't look the same, and you're stuck. That rain and darkness hid a passel of faults.

The same thing happens with real estate. When that piece of ground is covered with shrubs and trees and lush grass and flowers, it's beautiful. The defects are covered. But when the leaves fall, you have a marvelously clear view of the hog pens across the way. And the traffic noise surrounds you, unmuffled. Most of the ground is sticky clay mud, instead of lush pasture grass. And there aren't nearly as many evergreens as you thought. The winter winds roar straight across from the fertilizer plant.

That's a hassle that just makes no sense to me, especially after watching a number of folks struggle through it. Your best bet is to pick the areas you want to investigate, and when middle or late fall arrives, start looking in earnest. With or without a real estate agent. When the leaves are down, you can see what you're looking at—and find what you're looking for. You can tell tree sizes and shapes and conditions, spot the evergreens, determine the view planes, and see what lies beyond the summer leaves. You can find the boulders and outcrops, note the soil types, see the natural contours clearly, and determine the sun path and wind directions easily. If the lot is a poor one, you will know it immediately. And if the lot is a good one and has features that attract you, rest assured that it will not change for the worse in the summer.

And incidentally, unlike car shopping, a rainy day is perfect for lot hunting. Put on your boots and slicker and go to it. There won't be anybody else about to bother you. Furthermore, the gray grimness of a rainy day does three positive things for you. You are likely to be less exuberant, more practical, with a prove-it-to-me air about you. Second, a soaked landscape always seems more sharply delineated, and all the earthy tones and colors are intensified. You really see better what you are looking at. And third, for most people, rain tends to put everything in the worst possible perspective. If you like what you see under those circumstances, then very likely nothing will alter that opinion under better circumstances.

If the rainstorm is a substantial one that has continued for several hours before you go out scouting, you can pick up more useful facts. You can determine the storm direction, which probably will be the prevailing wind direction at that time of year. And with that information you can figure out what trees act as windbreaks, what ridges or rocks are weather shelters, and what spot might be best for a house under adverse weather conditions. You can also check to see where all that rainwater disappears to. Are there mushy spots where the ground is holding lots of water? Perhaps that particular soil has a poor absorption rate and is soaking the rain up too slowly. Or there might be gullies, ravines, small washes, or other runoff spots that have to be diverted or drain-tiled in the interest of dry foundations. If the site is under serious consideration, it wouldn't hurt to make notes or sketches, or both. This will give you a bit more data to work with when the time rolls around for the final decision.

There's another point in favor of fall lot shopping. The fall is the off-season for real estate agents. So the real estate agent will have time to talk, to discuss, to ponder your needs, to trail around with you at length. He may be hard up for a sale too; you could be his last chance for a commission before next spring. Your banker will have time to talk, too. You can get your name right at the top of the excavator's list; he can start digging just as soon as the frost is gone. The general contractor and his subs will be receptive because the fall is when they start trying to line up work for the next year. They'll have time to go over the details of your house with you well before starting time.

Now, as to the time for building. Spring is indeed a good time to start, unless you happen to be in one of the arid areas of the country where winter never really happens and you can begin anytime. But the spring immediately following a land purchase may not be so good. If you're not in a tearing hurry and can restrain yourself for another year, you will gain some advantages. There will be plenty of time to design your home, live with those plans for a while, and make adjustments as new ideas occur. There will be plenty of time to find and consult with contractors, engineers, architects, officials, etc. And there will be time to become familiar with your new lot so you can make the most out of its features as you design.

These advantages may not sound like much, but they are. You'll have fewer frustrations and fewer problems because you won't have to do everything at once. And that will ultimately result in a better, more livable, more enjoyable, more efficient home for you.

2

Getting the General Idea

With the preliminaries on the homesite out of the way, we can now establish some of the basic design parameters for the home itself. To do this successfully, you'll have to do some soul searching. You'll have to figure out exactly what you want *and* need.

Some years ago I was acquainted with an eminently successful architect who designed nothing but single-family dwellings. His fees were steep, his designs imaginative, his houses expensive. But he was so busy that he had to turn down far more commissions than be accepted. Why? Because he had a reputation for designing houses that worked, that became *homes*. And his clients were always happy and well satisfied. Their home living requirements were invariably fulfilled to a high degree.

This gentleman operated on two fundamental principles. The first says that the home should be an integral part of its environment. The whole should be a blend of its parts and should in turn meld with the surroundings. The second theory says that the home should not only be an extension of the personality of the owner, but a reflection of the personalities of all who live therein. The reflection shows in the materials used, the textures, the shapes, and the colors. And the reason that this architect's homes were so successful is that he took the time and trouble to adhere to these two principles.

And that's what you have to do, too. You have to insure that the home you design evolves out of a consideration of your own needs and desires. And you have to think hard about how your home can fit in with its surroundings.

LISTING YOUR REQUIREMENTS

Once again, you'll want to make some lists. Start with some hard facts about what you need and want in a house. Here's a sample list:

> Item: House guests. Mother-in-law visits four times yearly, brother once, friends once. Need one guest room big enough for two.
>
> Item: Cooking. Nearly all meals will be eaten at home. Need well-equipped, workable, efficient kitchen.
>
> Item: Eating. Need adequate space to feed three in comfort.
>
> Item: Relaxing. Need living room. Should be large enough for fireplace.
>
> Item: Facilities. Must have at least one bath.
>
> Item: Garage. Must have garage space for at least one vehicle.
>
> Item: Watch sports on TV a lot, sometimes with friends. Need a TV viewing area.
>
> Item: Often have guests in for dinner and drinks. Should have dining area or dining room large enough to accommodate 10 people.
>
> Item: Living room should also be large enough to seat 10 people comfortably, and to hold 25 for stand-up cocktail parties.
>
> Item: Should have fairly large rec room.
>
> Item: Should have swimming pool, but not too large.
>
> Item: Need patio area and barbecue, lawn large and level enough for croquet layout, and place for horseshoe pits.
>
> Item: Need storage space for photography equipment and darkroom for processing.
>
> Item: Need to make garage large enough to house two antique cars.
>
> Item: Need shop area in garage.

Item: Kitchen must be large enough to house plenty of gear and materials.

Item: Need sewing area.

Item: Need private office with room for files and large desk.

Item: Need small work area in quiet spot with desk.

Item: Plan for two more children. Nursery must be convertible to bedroom.

Item: Put up freestanding greenhouse later.

Item: Furniture-making might become money-making avocation. Allow room for large woodworking shop or make original one expandable.

This business of planning, or at least making some allowances, for future expansion of the home is one that often falls by the wayside. Obviously you cannot know right now exactly what changes in the home will be required. What you can do is be aware that unquestionably some changes or additions will be necessary eventually and set up your current plans so that you won't box yourself in. At least arrange your house design so you can alter or add on without ruining the lines and functions of the original house, and without tearing half the original structure down.

So according to the above list, the proposed house will consist of three bedrooms, bath, ample kitchen, living room, dining room or area, rec room, activities area, darkroom, and small office. There will also be a three-car garage with shop, a swimming pool, and a patio area.

Keep your list handy, and as you go along, make any changes that seem desirable. The design of a home should always remain flexible, even in the final stages.

CONSIDERING THE POSSIBILITIES

Many professional planners believe that a house should be designed from the outside in. That is, the structural framework should come first; interior components can be worked out later. And, indeed, this approach is usually fruitful.

Deciding on a structural "style" can be a real challenge. But all you need is a starting point. Just something to work with, something that for the moment at least will approximate your idea of what you'd like. Pick a house style or a house shape. You may even want to combine parts of several, or features of several, or details of several, when the time comes. And after you've chosen that initial factor, then you can adjust that to suit both exterior requirements and interior needs and functions. That's not as complicated as it sounds.

If you are considering styles, there are quite a number from which to choose. One of the simplest is the shed-roof variety (Fig. 2-1). This is easy to build, presents a pleasing appearance in either small or large sizes, allows a great deal of interior flexibility, can have two floors, and is readily expandable.

Another popular style is the Cape Cod (Fig. 2-2). This too is flexible and can be expanded in a number of ways but is more limited than the shed-roof style. A Cape-and-a-half has an extra full floor. And the partial attic of a standard Cape can be made livable by the addition of dormers (Fig. 2-3).

If you extend to the rear of a Cape and change the roof line, you have a saltbox (Fig. 2-4). This can double the first floor area and add substantially to the space in the upper story. As with the Cape, dormers can be added in the front, and ells to the sides.

Another popular design, especially for vacation homes, is the A-frame (Fig. 2-5). These can be easily built in all sizes, with upper lofts, second stories, and full basements.

Fig. 2-1. The shed roof style of home is versatile, economical, and easy to construct. It lends itself nicely to modular building.

Fig. 2-2. The Cape Cod home is one of the most popular in the country.

Then there are the unconventional styles, such as the geodesic dome (Fig. 2-6). Prefabricated factory-built panels can be built up into a dome in a matter of hours. They make attractive and highly livable houses, ranging in sizes from 500 to 5000 square feet.

There are other styles of houses, of course, and there are also hundreds of styles which have no particular designation except for a fancy name assigned to them by the designer. Looking through a few of the many magazines and books on

Fig. 2-3. Living space can be easily added to a Cape Cod style by the addition of ells and dormers.

homes and home design is an excellent way to define for yourself what your home should look like from the outside, and what features you might enjoy on the inside.

It may well be that architectural styles do little for you. In that case, think about shapes. Or combinations of shapes. Cubes, cylinders, pyramids, circles, parallelograms, ovoids. Combinations of various shapes, or the repetitive use of one shape, can lead to some pleasing and workable designs.

Try drawing a floor plan outline in an interesting shape. Never mind about drawing interior walls. Try a circle, a triangle, a hexagon. Or you can try less conventional shapes: a cross, or a doughnut, or a boomerang. You can also use more than one of each shape, or combine different shapes into patterns that give you a pleasing effect.

Once you have a floor plan outline worked out, then you can decide how the walls (exterior) should be stacked up. They might be straight up-and-down or slanted in or out. You might add a second story on the whole affair or only on some part. Then you can put on a roof, and this could be flat, or butterfly, or gently pitched, or sharply peaked. It could even be scalloped or domed.

This sort of planning takes a lot of doodling and dreaming, but if you want a home that is totally different and highly individualistic, this is the way to get one.

As you can see, what we're talking about here is merely the shell, the envelope of the house, without regard to what

Fig. 2-4. The saltbox style is also a popular one and is easy to design and build.

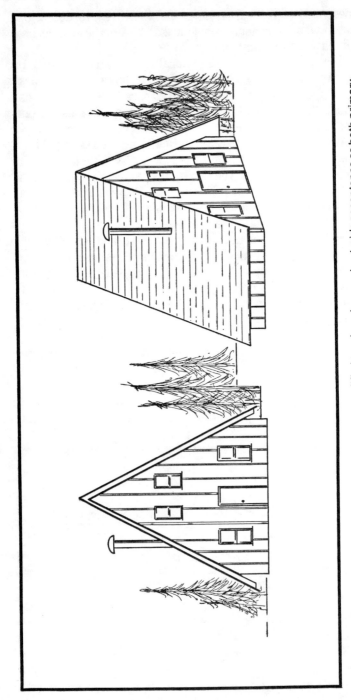

Fig. 2-5. The A-frame style and its many modified versions have gained wide acceptance as both primary and vacation homes.

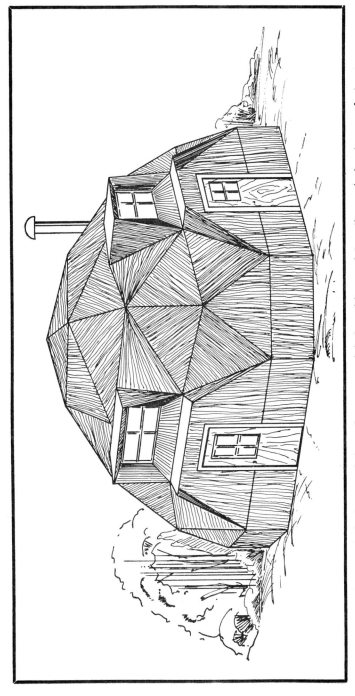

Fig. 2-6. The geodesic dome style home is an unusual design with many interesting and advantageous features.

may be on the inside. This exterior is your face to the outside world and as such is every bit as important as the interior. So before you get too far along in determining what type, style, or shape that envelope should be, let's look at some factors which will play a part in your final decisions.

First, how much living space do you need? In the initial stages of the design, working out both minimum and maximum figures can be a help since they will give you some guidelines to work within.

The simplest way to sort this out, in a general fashion, is by comparison, especially if you have trouble with spatial relationships or dimensional realities. Go back to your list of basics and attack the problem room by room. For instance, you know you want a large master bedroom. Let's assume that the one you have now is too small. How big is it? You measure the room and find that it is 10 feet wide and 14 feet long. How much bigger should it be ? Well, let's say your present living room is 14 feet wide and 20 feet long, and that looks as though it would be about right. Close your eyes and visualize your living room converted to a master bedroom with all your furnishings in place. If you are satisfied that this would be a good size, give or take a little, make a note that the master bedroom should be at least 280 square feet. That's what we're after right now, the square footage. And what about the maximum, speaking practically? Maybe, you think, 20 by 20. Anything larger you feel would be ridiculous. That would be 400 square feet; make a note.

Go through all your requirements this way, establishing the figures by comparison with something you live with or can actually look at and think about. Make one list for your present needs, another for the future, and separate the living quarters from such things as workshops and patios. When you are finished, you'll have a list that looks something like this:

PRESENT (in square feet)

Room	Min.	Max.
Master bedroom	280	400
Child's bedroom	140	200
Bath	60	120
Kitchen	128	200
Living room	384	600
Dining room	144	200

Room	Min.	Max.
Rec room	800	1000
Activities area	150	300
Darkroom	100	200
Office	100	200
Shop	400	600
Garage	576	720
Patio	400	?
Pool area	800	1500
Decks	?	?

FUTURE (in square feet)

Nursery	200	260
Bedroom	140	200
Studio	256	300
Shop expansion	200	400

When you boil all of this down, you can see that you will need a minimum of 2286 square feet of living quarters, or a maximum of 3420 square feet. That could mean the entire amount on one floor, or some combination of footages on various levels.

In addition, you need to find a spot to put the garage and maybe a patio or some decking or a pool or anything else you want.

One factor that will undoubtedly influence the size of your dream house is building costs. Let's assume that building costs in your area run $38 per square foot for standard frame construction, no frills. That would mean that 2286 square feet would cost about $87,000; 3420 square feet would cost about $130,000. If estimated building costs are too steep for the size you want, you'll have to scale down your dream house or somehow reduce construction costs.

After you determine the square footage, the next step is to relate the house size to your homesite. Will everything fit on the chosen spot? Perhaps if you want to put everything on one level, the layout will sprawl too far, run afoul of some big trees that you want to save, hang over the edge of the knoll, or run into a ledge. Maybe you should think in terms of two floors, or multiple levels, or a combination of connected units or pods instead of a monolithic structure. One way or another, you'll have to arrange your design so that everything fits upon your site comfortably. Preferably without any expensive site alteration. If your plans don't fit in well at this stage, you have two choices: Alter the plans or change sites.

The next question to ask yourself is whether your proposed design will blend into the surroundings? And most of the decision will be based on your personal feelings. For instance, you may decide that your favorite style, the Cape Cod, somehow looks out of place in the middle of 5 acres of sagebrush in Colorado. In some areas, a cubistic modern fits nicely; in others, a log cabin looks entirely appropriate.

The usual inclination is to make the structure merge with and complement the immediate surroundings, and vice versa. The log cabin is a natural for a wooded grove. A-frames and Swiss chalets blend in nicely with wooded or mountainous spots.

Of course, you may not care a hoot whether the structure fits the surroundings or not. You might even enjoy invading the neighborhood with something shockingly different that will have all the neighbors buzzing. But you have to approach this with some moderation and judgement. You've got to look at the resale potential in later years. The fact is, while many people ooh and aah over distinctive and unusual homes, they don't want to buy them. They'll stick with the conventional numbers. Not everybody feels this way, mind you, but enough do so that a conservative home in a conservative area will sell much faster and for a higher price than something highly unusual or outlandish in the same area.

When you're planning your house, you have to think about building codes, too. In some places, for instance, you simply can't get a permit to build a log cabin, no matter how nice it might be. Or you might want to build a saltbox. A saltbox should by rights have a cedar shingle or shake roof to look right. But there may be a local ordinance prohibiting the use of wood as a roofing material.

Figuring a way around such restrictions may take some careful thought and planning. If you opt for a traditional house style, you may have to consider radical design variations.

You may want to consider designing your house with different levels. There are two different types: levels within the structure and levels of the structure itself. A loft area extending out over a single-floor living area is one example of the former; the familiar tri-level home is an example of the latter.

There is something to be said for building with multiple ground levels. This makes it possible to use sloping ground effectively with a minimum of site preparation and ground disruption, and also leads to some exciting designs with great visual impact and exceptional harmony with the surroundings (Fig. 2-7). This can also be an expensive way to build, but sometimes, depending upon what you like and don't like, the results justify the costs. An extreme example of multiple-level design is what I call the tipple style (Fig. 2-8). Figure 2-9 shows another possibility, using conventional shapes and structures in small modules connected with stairs and passages.

Interior levels can also be effective and can be built rather inexpensively. Such levels can be completely open, partly walled, or closed off. And they can actually constitute a partial second floor with less of the building problems and costs of a second floor.

TYPES OF CONSTRUCTION

Now let's look at the different types of construction used in building houses. By far the most common is called *frame construction*. In this method, dimension stock (2×4s, 2×6s, 2×8s, etc.) is put together to form a structural skeleton. Floor joists and a sill are laid upon a level foundation of

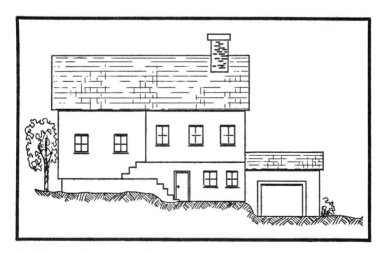

Fig. 2-7. A split-, tri-, or multiple-level home often has design advantages on uneven sites.

Fig. 2-8. Patterned after a mining tipple or stamp mill, this home design takes full advantage of a sloping site.

concrete or masonry, and a subfloor is laid upon that (Fig. 2-10). Then stud walls are erected, usually of 2 × 4s. More joists are laid across the tops of the walls and interior partitions, and then the roof rafters or trusses are set. The exterior skin of siding and roofing is put on, then an interior skin of drywall or plaster, and then windows, doors, and trim inside and out.

Post-and-beam construction also employs a wooden skeleton, but in a somewhat different fashion (Fig. 2-11). This is a traditional method, by which most of the old homes in the country were built. The resulting structure is exceptionally strong and sturdy, if properly done, and a great many of those houses built 200 to 300 years ago are still in use and good repair. Heavy wooden beams, usually 6 × 8s or 8 × 8s, are laid on top of a foundation to form the sill. Then a number of posts, of approximately the same size as the sill beams, are stood up on and secured to the sill. Beams are laid across the tops of the posts to join them, and directly above them more beams are run to form principal rafters. The wall spaces in between are then studded or crossbeamed and joists are set in the floor openings. The roof may be finished out with either

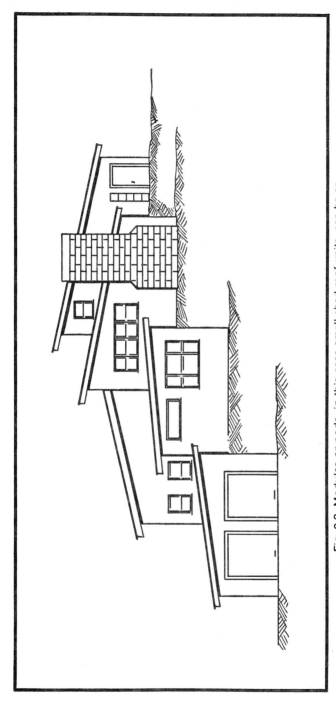

Fig. 2-9. Modules or pods, in this case small shed roof units, can be arranged and connected at different levels with minimum site disruption on an uneven lot.

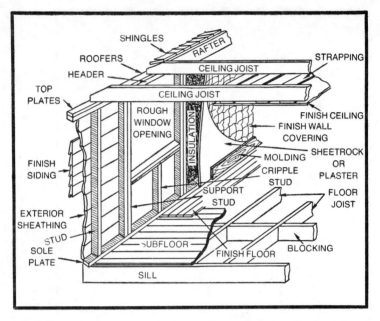

Fig. 2-10. This cutaway sketch shows the components of basic platform-framed construction.

additional common rafters and a skin of planking, or a series of longitudinal purlins across which the roofers, or roofing planks, are laid in up-and-down fashion, lying with the slope of the roof. Then the remaining inside and outside skins are applied in more or less usual fashion.

Pole construction joins the house and the foundations into one unit (Fig. 2-12). In this method, thick poles of suitable height are planted directly into the ground, just as a telephone pole is. The house structure is then built up around the poles, suspended from and supported by them, using standard wood dimension stock and conventional carpentry. The poles are sometimes exposed on the outside of the building as a part of the design, or they may be concealed within the structure.

Solid masonry construction is a type you are probably familiar with. Stone has been used for this purpose ever since man learned how to stack them up; brick was a later innovation. Today various types of concrete or cinder block may be used, either alone or in conjunction with brick and stone. Adobe is another material which could be placed in this class. High costs for labor and material are drawbacks to constructing masonry homes these days, but many are built every year (Fig. 2-13).

More common are the masonry veneer types of homes. The basic structure is built in much the same way as a frame house, but all or part of the exterior is covered with a thin layer, or veneer, of brick or stone.

There is another type of construction made possible by the use of modern-day methods and materials, one which is rather

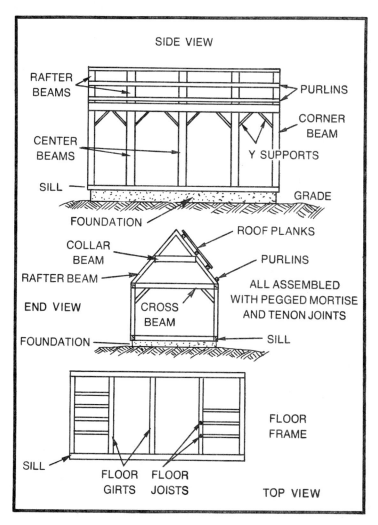

Fig. 2-11. The old method of post-and-beam construction can be used in buildings of almost any size or configuration.

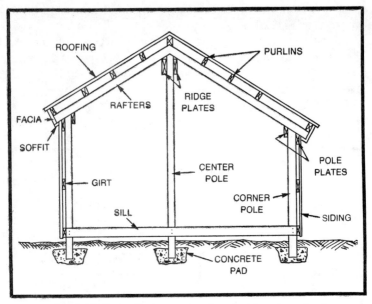

Fig. 2-12. Typical pole construction, inside-pole method.

unusual and not in widespread use but which certainly makes a unique structure. Details vary depending upon just what sort of materials are being used, but essentially this is a *free-form construction*. A framework is built of metal tubing or some other rigid material and then covered with a wire mesh. The configurations can be nearly any combination of forms and shapes: domes, slants, peaks, etc., all flowing smoothly together in whatever arrangement suits the whims of the designer. This skeleton is then covered by spraying on a semiliquid plastic material, inside and out, which soon hardens, or cures, to make a rigid structure.

As you can readily see, there are lots of possibilities for you to investigate and mull over in establishing the basic design parameters for your house. And perhaps one of the most important aspects of home design is the realization that there *are* so many possibilities, so many different ways that you can go about accomplishing the end result. Though there is a lot of work and a lot of head-scratching involved, there is no substitute at this early state of the game for digging up all the information you can. Look through periodicals which deal with home design and home construction, check your library for books which deal with architecture and construction. You'll pick up a lot of valuable ideas, as well as plenty of solid knowledge on how houses are actually put together. Examine your friends' houses, drive around the neighborhood looking at more houses, even stop to watch a few homes being built if you can. You may get dizzy with details for a while, but the more input you have, the more relevant your final decisions will be, and the easier this engrossing chore of designing becomes.

HEAT, LIGHT, SHADE, AND WEATHER

There are four items of great importance to every household: light, weather, heat, and shade. All have a substantial impact on your daily household living. You can maximize or minimize those impacts, or you can disregard them. The latter course generally results in higher operating expenses and more than a few irritations later on.

If you have been reading anything about solar heating lately, you know that the position of a building relative to the sun can be very important in a number of ways.

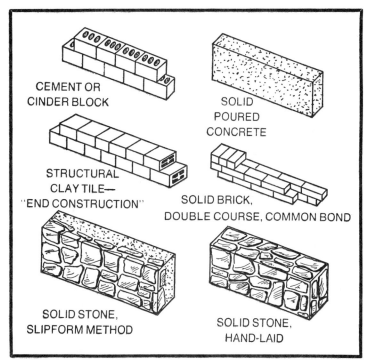

Fig. 2-13. Some of the many materials for solid masonry walls.

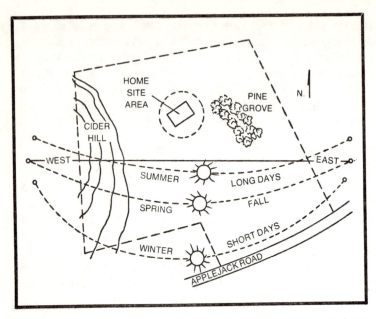

Fig 2-14. If you want sunshine on your shoulder, you'll have to know where to find it. Coordinating the seasonal sun tracks with your home site and building location will give you some answers.

Every schoolboy knows that the sun rises in the east and sets in the west. But the position of sunrise and sunset changes, the angle of the sun changes, and the length of time the sun shines varies depending upon the time of year. The easiest way to get an idea of how sunlight strikes your lot is to make a few simple observations, which you can then translate into a sketch like the one in Fig. 2-14.

First draw a rough map of your lot. Scale is not necessary, just so the proportions are about right. Take the sketch to your homesite, along with a compass, and orient the map. You can do this by lining up one of the sketch boundaries with the actual boundary. With the two lined up, set your compass on top of the sketch, away from any metallic objects (like belt buckles or car hoods) which might influence the compass reading. Adjust the compass until the needle and the N on the compass dial are both aligned. Then draw in a north-pointing arrow on the sketch. Now draw an east-west axis through the plot sketch.

When the sun is in the equinox position, and the days and nights are of equal length (twice a year), the sun will indeed rise in the east and set in the west. But during the summer,

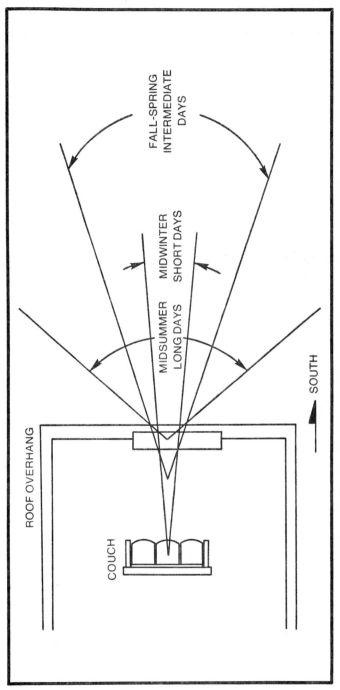

Fig. 2-15. The length of time that the sun strikes any given spot also shifts about.

sunrise will be in the northeast and sunset in the northwest. By wintertime, sunrise will be in the southeast and sunset in the southwest. What this means is that each day the sun will shine in through your windows or down into your greenhouse at slightly different angles. For instance, in the summer, a couch in my living room is in full sun from about 6:54 a.m. until almost noon. By early fall, the time has diminished to around 7:30 a.m. to 9:30 a.m., and the strike angle is much shallower (Fig. 2-15).

That shallower angle occurs because there is also a difference in the height of the sun from season to season (Fig. 2-16). In the summer the sun is at its highest, shining down at its steepest angle with respect to the earth's surface. In the winter the sun is lowest, and the angle shallowest. In spring and fall, the midpoint is reached. One easy way to measure this angle is to plant a stick in the ground when the sun is at its highest point for the day, or approximately so, around noontime. Adjust the stick so no shadow is cast. Set a level horizontally at the base of the stick and adjust it so the bubble

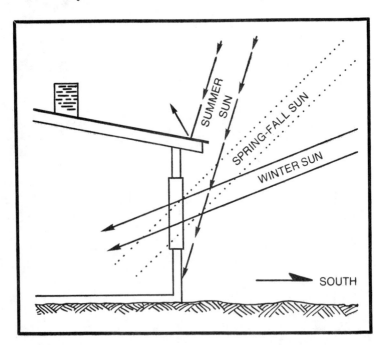

Fig. 2-16. The angle of the sun relative to the ground changes markedly through the year, with a varying impact upon your home.

is centered. Then measure the angle between the level and the stick with an ordinary protractor.

You should also make note of the length of time that the sun shines on your particular site. Just find out when the first rays of sun fall upon a point about where your house is likely to be, and when the same point goes into shadow. Of course, this period of sunlight will be longest during the summer and shortest during the winter. By consulting tables from the nearest weather bureau or *The Old Farmer's Almanac*, you can find out what the approximate sunrise and sunset times are for your general area. By comparing this official information with your own observations from your homesite, you can get an approximation of how long the sun will shine down upon you at any given time of the year.

Now, how does all this affect your home design? In several ways. First, heat. If you are planning to use solar comfort heating in your home or solar water heating for the domestic hot water supply, you will need all this information along with a lot more. The solar collector units must be positioned so they are exposed to the rays of the sun as long as possible during each day and for as many days as possible, especially during the weeks when the coldest weather can be expected in your area. So the collectors must face due south, and any deviation will lessen maximum efficiency, unless the unit happens to be automated to move with the sun track.

If the collectors are roof mounted, the roof, or the collector support system, and consequently part of the house itself, must face south. All this may make a difference in the siting or design of the house.

Whether you plan on having a solar heating system or not, you can make good use of the sun's rays in heating your home. The more sunlight you can get into your home, the more free heat will be introduced and the less often the thermostats will click on. Here you must consider the room plan, the siting of the home, and the use of the rooms which will receive sun. For instance, let's say you want your large living room, with its large windows, at the front of the house. These large windows could allow the sun to keep the room warm all day, even though there is a lot of interior volume to be heated. But to do this, the largest glass area should face due south, or nearly so. The view that you so dearly love, however, may be to the north. Now what? Compromise. Try to reposition and redesign so that you can get plenty of sun and still see the view. It can be done.

If you know the angles of the sun at various times of the year, you can adjust the angle of the roof, the height of the windows, and the length of the roof overhang to get just the amount of sun that you want.

For instance, in the summer you don't usually need any extra warmth from the sun; you've got aplenty and probably too much. But this is when the sun is highest and the rays are at the steepest angle. Deep eaves or long roof overhangs plus the right roof pitch will keep the sun from shining in too much. But in the winter when you welcome that warmth, if the roof pitch and overhang has been figured correctly you will get plenty of sunshine beaming right in (Fig. 2-16). Protection for furnishings, upholstery, and carpeting from the deleterious effects of direct sunlight can be provided by drapes and blinds, as well as judicious placement of furnishings and use of sun-resistant materials within the house.

Light coming in from the east and west, at sunrise and sunset, is the warmest in tone. North light, especially from a deep blue northern sky, is the coldest. South light is the brightest, for the most part, and a house which is open to a wide sweep from east through south to west will receive the greatest amount of light for the longest period of time through the day. Northerly lighting is shorter in duration and of less intensity.

And then there is the converse, shade. Much as we like and use sunlight, we also enjoy shade. By using this same information about how the sun strikes your property, you can also determine where the shade will be, and when, and for how long. Such information can play a part in the location of the patio, the barbecue pit, or even a portion of the house (Fig. 2-17). You might feel, for instance, that having one side of the living room in deep shade during the latter part of the summer afternoons outweighs the advantages of additional sunlight during the winter months. And if you are planning to have decks, you may be able to arrange them so that some sections will have sun and others will be in the shade, no matter what the time of day. You might not like the early morning sun coming into the bedroom. Situate the bedrooms so that a later sun comes in, or none at all (north side). You might enjoy a sunny and cheerful kitchen at breakfast time. And if the

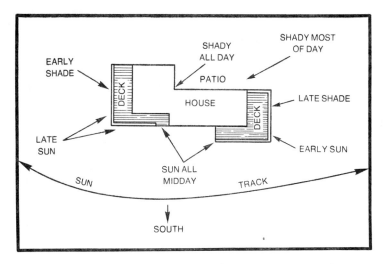

Fig. 2-17. With proper planning, you can arrange the house and grounds so you will have either sun or shade available where you want it.

morning sun enters the kitchen, then that low, hot summer afternoon sun will not, which makes the task of cooking meals in a hot kitchen a bit less bothersome. If you live in a hot, arid area where winter really never happens, you can build a low-walled one-story home with long roof overhangs and position the building with the narrowest end facing south (Fig. 2-18). This will lower the impact of the sun's rays on the entire building and reduce cooling costs and problems.

Sunlight is free and can work for you in many ways. So keep the sun track in mind all the time, in all its seasonal variations, as you lay out your estate.

The other item of major importance in siting a house is the weather. And though the underlying weather patterns in this country for the most part travel from west to east, there will be plenty of regional and local variations. The only way to tell with certainty is to observe what happens on your own parcel of land. Even such topographical items as a tall ledge, a ridge, a grove of trees, or even a nearby large building can play a part.

Usually there will be one principal direction from which the wind will blow across your property. This direction may vary one or two points (a point equals 11 1/2°) of the compass from time to time. In my area, for instance, the principal wind direction is out of the west at 270°. The variation may run from

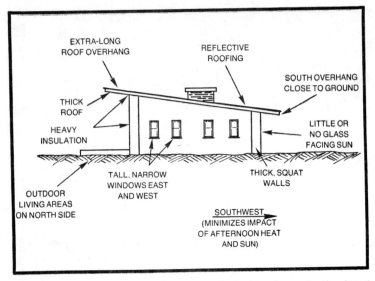

Fig. 2-18. There are many possibilities in designing a home for the desert country. Most major features are planned to minimize the impact of the sun and sometimes wind as well.

about 250°, a bit south of west, to 300°, a bit north of west. Virtually all of the stormy weather follows this same track.

There may also be a secondary direction from which gentler winds and lesser storms blow on a less frequent basis. Where I am, this secondary weather pattern comes out of the east and usually varies only a few degrees one way or the other. In fair weather these winds are light breezes, and in foul weather the storms are generally light thundershowers or snowsqualls.

There is also the possibility, depending upon the terrain that surrounds your homesite, that there will be fair-weather breezes from one direction during the day, with the air flow reversing itself during the nighttime and perhaps part of the early morning. The day breeze will be warm and flow uphill, while the night breeze is caused by cold air flowing downhill. This is a situation familiar to campers and mountain climbers who frequently encounter these thermal air currents. Air temperature differentials are the cause, and they are not a part of the normal prevailing wind patterns, though they may be influenced by those patterns.

Once you determine what the wind/weather directions are at your location, you can put the information to work for you.

First, you can position the house so that the impact of the wind is minimized. The smallest areas of siding and roofing should be set into the wind flow. This minimizes the weathering effects upon the building, decreases the possibilities of damage during severe storms, and helps to reduce heat loss from the house during cold and windy weather.

Another possibility is to place the back of the house (which can also be the smallest or narrowest portion) to the weather. This puts storm tumult and inevitable interior cold air currents away from the areas of the house that are in general daytime use.

The weather side of the building should also have as little glass as possible, for the same reasons outlined above. Exterior doors can be placed on the lee side, which makes entering and exiting somewhat easier and helps to keep cold wind and precipitation outside where it belongs. Or, doors on the sides of the building can be hinged to open so that they form a shield for the user, with wind pressure operating to shut them rather than wrenching them open. A door hinged against the wind flow acts as a scoop, funneling air and moisture into the house, and if the user loses his grasp, the door could be ripped off its hinges.

Windows can also be arranged to provide a breezy crossventilation. Sliding and double-hung windows can be set so they are at angles to the wind flow around the house, picking up only a portion of the passing breeze. Casement windows should be hinged so that when closed, the air flows right on by, but when opened into the air flow, they will act as scoops to channel air into the room. By carefully figuring the air flow you can, with some compromises, arrange the window openings so you can get good crossventilation in any room in the house. Tricky, but possible.

Developing Preliminary Plans

After you get some general ideas about the kind of house you want, you can get down to some specific planning.

THE SITE SKETCH

Every design is different, naturally, but the general process of developing the design is similar. So in order to illustrate as we go along, I'll work up a hypothetical homestead. The piece of ground I have chosen is shown in Fig. 3-1. The storm track runs generally out of the west. There is a brook close at hand, and the road is far enough away so that I won't be bothered by traffic noise. There are some big trees; the site is relatively flat; there is plenty of room.

So the first step is to work up the site sketch, like the one in Fig. 3-2. All the major elements are in place, roughly, and the orientation of the building with respect to those elements is now fairly obvious. But not rigidly set. Never that. The key to workable home design is flexibility. I know that I want to use a random shape to begin with, maybe a combination of shapes, but I don't know yet exactly what. I want to make the structure fit into the surroundings as closely as possible, with an absolute minimum of site preparation, even if that means a little extra care and time at the outset. So for the moment the main house is simply represented by a rectangle.

Once the site sketch is made up, go back to the actual site and look around once more, this time with a 50-foot tape in hand. Pick some definite spots for the house, garage, pool, etc., set up some rough outlines, and check the distances between the various parts of the homestead. Now, should the house face a bit more this way, or that? Should the drive go around the other side of that big rock, or this side? Perhaps by turning the main structure a bit you can see more of the view. Maybe, now that you look again, the structure should be long and narrow instead of nearly square, or if you make it curved it will fit in better. New ideas may present themselves, old ones might not look so good the second or third time around.

Don't be discouraged by this. Welcome it. No harm can possibly come from looking, then pondering, then looking again. Keep the process up until you are well satisfied and easy in your mind.

After you know the site like the palm of your hand and have the site sketch to refresh your memory, go to work on the house floor plan.

THE FLOOR PLAN

A floor plan is a drawing that shows the relative location and sizes of the rooms within a structure, and usually the outside features like decks or porches as well. Permanent major items like doors, windows, fireplaces, and room

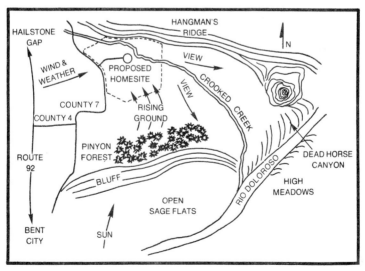

Fig. 3-1. This large-sized sketch shows the territory surrounding my hypothetical homesite.

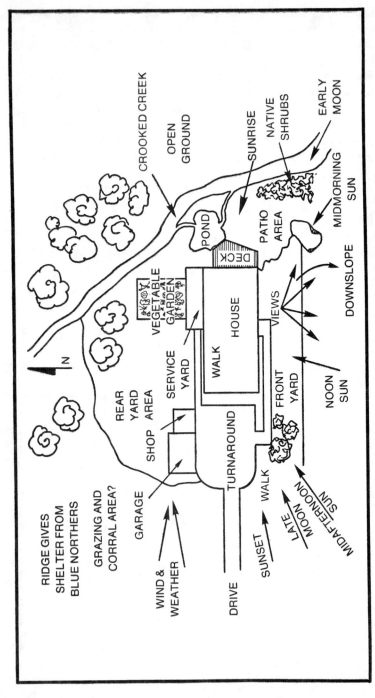

Fig. 3-2. This sketch is a rough plot plan of my hypothetical homesite. Here I have set my proposed home into the existing natural features of the site in a way I think best at the moment.

dividers are also shown. The plan is one-dimensional, and a separate plan is used for each floor or level. In a preliminary floor plan, only rough dimensions are used and the sizes and shapes are approximately proportional. Final floor plans are drawn to scale, showing dimensions and sizes to a certain specified tolerance. For the moment, all we need are rough sketches.

The starting point is the shape, the bare outline of the building. This will be the shape of the first floor, or ground floor level. The shapes of other floors or levels could well be different. The contours of my site, the terrain, the directions of view, the sun and weather tracks, and the position of the access road suggest to me the possibility of an A-shape, or perhaps a Y, or some modification of either. I think I can do more with a Y, so the beginnings of my floor plan will look like Fig. 3-3. Yours will unquestionably be considerably different.

Now I have to fit a certain number of rooms into the outline. I also have to arrange them in some logical fashion,

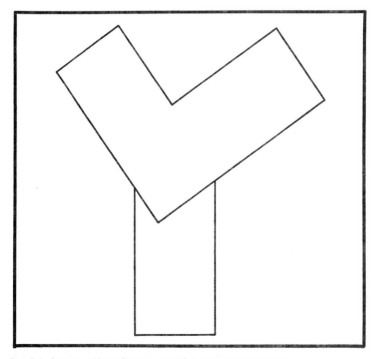

Fig. 3-3. A simple Y configuration will be the starting point of my house design.

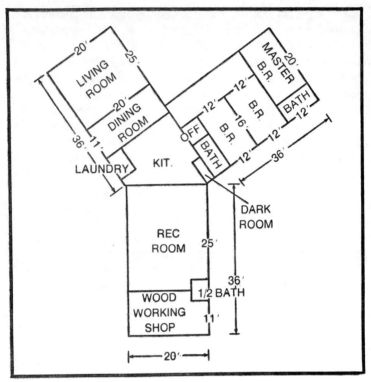

Fig. 3-4. The first rough sketch of the Y house floor plan.

provide good access to all of them, and establish decent proportions and sizes. I will need three bedrooms, a bath, a kitchen, a living room, a dining room or area, a rec room, a small office, and a woodworking shop. There must be a garage, too, but I'm not yet sure if I want that attached to the house, an integral part of the house, or freestanding. I'll put that decision off for a while.

First, the living room. I want it to be large but cozy, comfortable, and functional. The principal exposure should be toward the view. Twenty feet might be a good width, and perhaps 25 feet for the length. An obvious location would be in one leg of the Y, taking up nearly the entire leg (Fig. 3-4).

Bedrooms should be away from the action, where there is some peace and quiet. If they are well separated from the living room, there will be little interference between the two diverse use patterns. Three are needed, and the master bedroom I feel should be larger and more luxurious than the other two. The ideal spot for all three bedrooms would be in the other arm of the Y. But there's a problem. How do you stack three small rooms in a row within a long narrow structure and gain access to all three without going through one or another of the rooms?

Well, here's one solution. If the living room arm is 20 feet wide, then the other arm should be the same to maintain balance. So by putting the master bedroom across the tip of the second arm, that will make a room 20 feet wide. The depth we can arbitrarily set at 12 feet for the moment.

Now I can run a 4-foot-wide corridor down the inside wall of the arm. Off the corridor I can locate the remaining two bedrooms. Each will be 16 feet deep, and I can pick an arbitrary width of 12 feet.

The kitchen should be as centrally located as possible, for a number of reasons. For one thing, in most households the kitchen is the core of daily life. For another, it is the center of work activities which affect all other parts of the house. And for a third, it is essentially a utilities center.

This is where food is prepared, so ready access to the dining room is necessary. If there is a breakfast nook in the kitchen, this is where the family will automatically congregate first thing in the morning, so there is a bit more convenience in having the kitchen near the sleeping area.

And from the kitchen or its immediate environs come the cleaning supplies to administer to the remainder of the house. Since it is something of a work center, this is the general area where the power enters, the water supply comes in, and it is one of the principal points of liquid waste drainage and rubbish accumulation. The kitchen, then, should be in the center section of the Y for convenience all around.

Now the dining area. The logical spot for that is between the living room and the kitchen. There is a definite traffic pattern between all three rooms. Prepared food goes into the dining room; dirty dishes go back again. Guests go into the living room, then into the dining room, and back to the living room. Yet with the proper placement, the activities of any of the three rooms will not intrude upon the others, even with all three operating at once, as so often happens when the gang drops by on Saturday night. So I'll place the dining room in the first arm of the Y.

That leaves the tail of the Y. Since the two arms are each 20 feet wide, the tail might just as well be the same. The next room to position is the rec room, and this would be a perfect spot for a room which usually generates a lot of noise and activity. It is well away from the rest of the house and whatever else may be taking place. So the rec room can go next to the kitchen in the tail of the Y. For length, I can choose 25 feet as a good round figure.

The workshop can be a noisy and dusty spot, full of paint smells and shavings to drift around. I'll put that right out at the end of the Y. It will be 20 feet wide.

So far I don't have a length for any of the three arms of the Y. The three bedrooms total 36 feet though. I can make that the length of the inside of each arm of the Y. The living room takes up 25 feet of one arm, so I can make the dining room 11 feet long, equalizing the two arms. Seems logical that the tail should be the same, which would give me a 25-foot rec room and an 11-foot shop.

The bathroom is missing; that will never do. In fact, looking at the size and layout of the house so far, it seems obvious that only one bath, as originally planned, would be a mistake. There should be two, maybe two and a half. The trick with baths is to locate them as close to the kitchen, and as close to each other, as you possibly can. This saves a considerable amount in water and waste piping and tends to centralize the plumbing system. This also places the hot water supply within easy reach, so that hot water is more likely to be there when you want it. Such an arrangement minimizes the heat loss from the hot water system. However, in a sprawled-out house like this one, that ideal situation of a pair of back-to-back baths which are close to the kitchen and laundry may not be so easy to achieve. And of course if you don't care about the extra expenses, then the locations don't make all that much difference. Even that minor irritation of waiting for the water to run hot can be overcome, with the right equipment.

I like plenty of privacy in the bath without a gang of folks knocking on the door, so I'm going to have one full bath accessible only from within the master bedroom. It's unfortunate that this preference places the bath at the far end of the bedroom arm, but for the moment at least, that's the way it'll be. I'll just have to pay the added price for my foibles, unless I can devise something better.

Of course, there will have to be at least one other bath. The living room is no place for a bath, and having one off the dining room doesn't seem too good either. Somewhere around the kitchen would be all right. And if I place it to the right of the kitchen, then I can make it serve multiple duty. It will be handy to the kitchen, and it's fairly handy to the rec room. At the same time, I can cut an access through to that first bedroom. Which in turn means that this bedroom would be the logical one for a guest room.

While we're on the subject, should there be more facilities available? Perhaps so. For the small extra outlay in plumbing, fixtures, and space, sometimes a considerable amount of convenience can be gained. With my Y house, it would be handy to have a place to wash up right on the spot after doing some work in the shop. And there might be some need for water during a workshop project. So having a wash basin in the shop would be nice. By placing it between the shop and the rec room, I can make it serve both areas. And since I'm going to put in a wash basin, why not go ahead and install a toilet as well? I'll draw a half-bath into the sketch as a possibility and see how it works out later on.

There are still some small rooms needed, and there are some odd-shaped spots around the kitchen area which I can use. A small office could be fitted in next to the guest bath. A darkroom could go next to the kitchen where the plumbing and hot water supply is handy for connections. The room needn't be particularly large, so between the kitchen area and the guest bath would be an appropriate spot.

Most, though not all, households have need for a laundry. This too should be near the kitchen. Putting it near the kitchen cuts down on plumbing and drainage problems and allows you to attend to the laundry chores while working in the kitchen at the same time. There's a small spot remaining to the left of the kitchen, so the laundry can go there.

Of course, in drawing your own floor plan, you can't afford to be as off-the-cuff as I have here. You'll have to think about all the considerations of Chapter 2—building costs, building codes, maximum and minimum floor areas, etc.

Now that I have a working floor plan, I can start adding details to it, making it more and more precise.

Passageways create some of the traffic patterns in the house, so we have to scrutinize those to avoid or minimize conflicting patterns.

One of the most obvious traffic patterns involves the hallway which serves the bedrooms. Each bedroom must have a doorway, and there is really only one location for the doorway leading into the master bedroom: at the end of the hallway, centered. The other two bedrooms will be entered from the hall. So the doorways at this point can be placed in the interior hall wall. I'll put them close to the partition that separates the two rooms.

There is only one place for the doorway to the guest bath. And the guest bedroom must have a doorway near the guest bath. There are only two possibilities for the darkroom, since there are only two interior walls. I'll just pick one and see what happens.

There will probably be a lot of random traffic back and forth from living room to dining room. Maybe it would be best not to have a doorway there. Instead I'll leave out part of the partition wall, all the way from floor to ceiling. And since traffic will flow around the inner V of the Y, I'll leave out that section of the partition. Seems like a good idea to do the same with the dining room-kitchen partition too. Now the dining room is more of an area, and the living quarters are more open and spacious.

The rec room can only be reached by going through the kitchen, and maybe that isn't so good. Changes may have to be made there. For the moment, I'll put a doorway in the right side of the rec room wall. Entrance to the shop must be through the rec room and can be anywhere. And in order that the half bath can be reached from either room, I'll need a door in each end.

Now, the exterior doorways. There should be a more or less formal front door somewhere, and the most logical spot would seem to be the living room. I could put it close to the dining room partition, but the doorway probably shouldn't open right into the living room. That can let in blasts of cold air. That means some sort of an entryway, which I will add in.

Now I haven't considered the patio and the decks. There should be some sort of way out to them. A patio area would fit nicely between the arms of the Y, and access to it could be through the living room wall. I can put a deck off the end of the living room with another door there. While I'm at it, why shouldn't I have a deck off the master bedroom too? That would be rather pleasant and would also make the building symmetrical. And by extending both decks slightly beyond the inside walls of the V of the Y, I can go directly from decks to patio with the help of a step or two to change levels.

How about the back door? In most houses the back door is the one which gets the most use and generally is located around the working center of the house, the kitchen. Again, this doorway would be better if it did not open directly into the kitchen.

And then too there are always those odds and ends which have to be stored but easily and conveniently reached. Coats, hats, snow shovel, broom, overshoes, boots, dog leashes, heaven knows what-all. So how about adding a fairly large back hall? Or better yet, a mud room? This could even have a wash-up sink in it and serve as a general transitional room for the whole family to use when moving from outdoor activities to indoor living. Maybe something could be fitted in around the laundry side of the kitchen. And the mud room will then have to have an interior doorway to lead into the kitchen.

The workshop should probably have an outside doorway too, a bigger one than normal so that bulky material can be taken in with ease. And if I want to build a boat, I'll be able to get it out without tearing the wall down. I'll put a doorway right in the middle of the end of the Y.

I have access to the house from the front side of the Y, the end, and one of the legs. But there's no way to get in from the backyard. That means another exit on the opposite side of the kitchen, near the darkroom. And there may be a problem with getting out to the patio too. I'll have a barbecue pit out there and will do some entertaining and relaxing. That means being handy to the kitchen, and right now I have to go through the living room and the dining room first. I can solve this problem in one stroke by squaring off the angle in the V of the Y and putting a doorway in there.

Now let's tackle windows. At this point I don't much care about the size or type, or exactly where they will be placed, or how high or low in the wall, or any other details. All that will come later. For now, I just want to spot the places where I think there should be some glass.

In my hypothetical house, I want to be able to see out wherever there is anything to see. So I'll put windows in all the exterior walls. The dining room area can have two fairly large windows on the two exterior walls. This will give a nice balance and plenty of light. The laundry should have a small window for light and to let the damp air out. The rec room I'll fit out with two fairly large windows on each outside wall. The

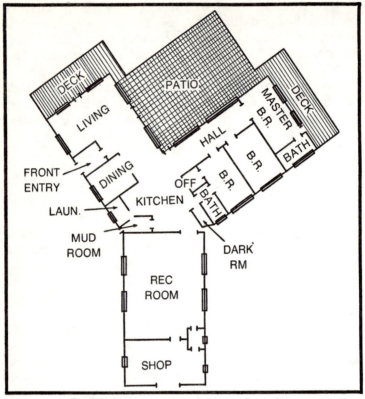

Fig. 3-5. The more detailed Y house floor plan.

shop gets one small window on each side. Even if I glassed all three shop walls, I'd still have to use a lot of artificial light for close work. Besides, I want lots of wall space to hang tools on.

The half bath should have a small window. Same with the guest bath. But the darkroom will have none. Each bedroom will get one fairly long and narrow window high on the back wall. The master bedroom, though, has some possibilities for a view and I'd like to take advantage of them. So I'll put lots of glass in that end wall.

How about the hallway? It's fairly long, and hallways always seem to be dark and dingy places. Let's get some daylight in there with a strip of glass high off the floor.

So there you have it. The completed preliminary floor plan is shown in Fig. 3-5. There are a lot of problems in there, and a thousand details to look after. But now there is something to work with, to shape, to modify, to rearrange.

PROVIDING FOR FUTURE EXPANSION

Not all houses are conveniently expandable. Some are really not expandable at all, without ruining the design or the general appearance. This means that the original house should be designed with the possibilities for expansion drawn right into the plans. The rule of thumb is that the least amount of the original building should be disturbed during the process of expanding. It is quite possible to remove the roof of an existing structure, add another story, and replace the roof. Or to cut a house in half, separate the two halves, and build up a new center section. Anything is possible. But the cost in time and money is tremendous in any major project of this sort, not to mention the enormous disruption in the lives of the inhabitants.

There are two ways to go. First, you can work out an open-ended design which lends itself nicely to additions as the years pass. Or you can build a large shell, with all the room you think you'll ever need at the outset, and then expand within the shell when you need to. Of the two methods, the former is by far the better. Rarely can you tell exactly what you'll require later on; ideas and life styles change.

When you work on an open-ended design, there are only a few items to keep in mind. First, make it possible to expand with the removal or rearrangement of as little material as possible. For instance, a smaller ell can be added to a Cape Cod house simply by removing some siding, providing a foundation for the new section, attaching the section to the original structure, and substituting a doorway for an existing window opening. Second, arrange the design so that as much of the old material removed from the existing structure can be used again in the addition. This would include doors and windows, possibly some of the structural material, and perhaps thermal insulation. Third, arrange the original design, or the design of the addition, or preferably both, so that one complements the other and neither detracts from the overall usefulness and appearance of the whole. And last, arrange the various accessory items, like a pool, patio, or garage, so that they won't have to be torn down to make way for the new addition. Leave plenty of room in which to grow.

Frequently these requirements are easy to meet. Sometimes they are not. About all you can do, really, is to make some reasonable choices and estimates now and hope that the future does not play too many tricks on you.

In providing for the future in my Y house, part of the process is easy, and part a bit more difficult. The greenhouse is no problem because I have plenty of space all around the building. Nor is the expansion of the garage difficult. Since I have decided to make that structure separate from the rest of the house, all I have to do is insure that there is sufficient space to add an extra bay onto one end.

Problems begin to pop up as I consider adding rooms to the main structure. The most obvious way to expand is to lengthen the arms. That would work with the woodworking shop, but the sleeping arm would require major rebuilding. I would have to make the master bedroom space over into a nursery plus an extra bedroom, and then move the master bedroom out another notch while making the hallway longer. And the deck would also have to be moved. And on top of that, I wanted to use a shed roof on each section, sloping down to each end. But since the roof is so low at the end wall, I would have to break the roof line because extending the roof at the same pitch would quickly bring it below ceiling level. I would have to tack on some small flat roofs, which wouldn't look right and would cause problems.

Well, then, perhaps I should rethink the design to allow for easy expansion. Is there some way I can simply add on to the ends of the arms and the tail of the Y without a whole lot of fuss and feathers? Yes.

First I'll have to change my ideas about the roof. If I really want to stay with the shed roof sloping from the center of the Y out to the ends of the arms, I'll have to elevate the whole roof by making the walls higher, and perhaps adjust the slope as well, so that after I add on later, the lower point of the roof at the new arm ends will be just above ceiling height. Until I do add on, the low ends of the roof will be 10 or 12 feet high, maybe more. And instead of putting in the decks now, I can let them wait until after the additions have been made. After all, I have the patio to use, and that won't be affected. The sleeping quarters will have to be rearranged somewhat, so that the hallway will extend all the way to the end of the arm, ready to continue into the new section to serve the additional bedroom and nursery. And none of these changes will be difficult or expensive, either in the present construction or in making the additions. Nor will there be any complications involved in living in the present structure.

Finalizing the Preliminary Plans

4

Your preliminary plans are just the beginning. There are many more items to consider before we can start the final sketching. And the first thing to do now is to finalize the preliminaries.

DOUBLECHECKING AT THE SITE

With your latest heap of notes and drawings in hand, go to the homesite once again. This time take along a bundle of stakes, a big ball of twine, and a 50-foot tape measure. The local lumberyard can supply you with regular survey or grade stakes, which are short, precut, and pointed for easy driving. But these are hard to see in brushy country or where there are numerous grade level changes. A bundle of plaster laths will do a better job because they are much longer. If you use lath, take some leather gloves along—the splinters in those things are murderous. You might as well organize a picnic too, and haul the whole family along. This next process usually takes some time and engenders all kinds of discussions. But if you are enthusiastic about your plans, it's fun too.

Get yourself properly oriented with the compass, and stake out all the corners of the house, measuring the lines and keeping the proportions and angles as true as you can. This may take a little restaking and fiddling around, but you can come pretty close. Run the twine around the stakes, and you

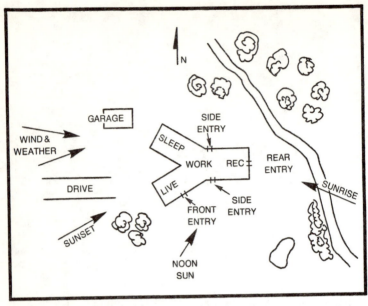

Fig. 4-1. With the plot plan redrawn and the Y house set into place, a number of problems show up.

will have an outline of the house. You should be able to visualize what the place will look like. With a few more stakes and strings, you can block in the main first-level rooms too. And now the question is, how does what you have lined out relate with your drawings, and how do they both relate with what you have in mind?

Chances are that you will discover some problems right away. Some of the rooms may seem a bit too small or maybe too large. The orientation of the whole building, or of some parts of it, may not be exactly right. Possibly you are not taking full advantage of the view, or a great boulder comes up through the deck. Maybe you should change the shape of an ell, move everything over 5 feet, reroute the driveway, put the front entrance somewhere else, cut down that big tree you had thought to save. Maybe there are some advantages in the site that you never even saw before, until you got those strings up and had something tangible to work with.

Whatever problems or new ideas crop up, make full notes on them. Then it's back to the old drawing board to make any necessary changes or adjustments and to add in the fresh thoughts.

Here are the results of an imaginary visit to my hypothetical homesite. I had planned to position the house with the tail of the Y pointing due east, but a second look shows that that won't work well (Fig. 4-1). Both front and rear entries are awkwardly placed. The V of the Y will get the full brunt of the storms and winds, trapping them and causing all kinds of strange turbulence, snow buildup, and odd structural pressures. The sleep section is close to the garage, a noise area.

If I turn the structure end for end, that will help (Fig. 4-2). Then the storms and wind will hit the narrow end of the building and deflect and disburse along the length of the building. Overall placement is better, and the sun track is still okay. No problem with the immediate natural surroundings. Now, by shifting the longitudinal axis of the structure from directly east-west to a few degrees northeast-southwest, I can align the building lengthwise almost exactly with the prevailing storm track and at the same time present one entire side or face of the building roughly to the south. This will give me shade and north light along most of the opposite long side.

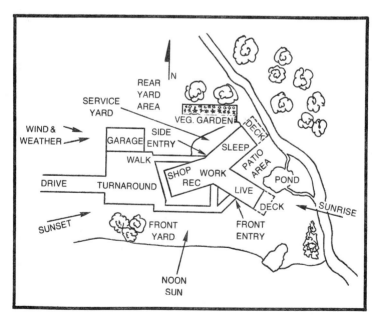

Fig. 4-2. Here the plot plan is redrawn, and the Y house situated differently. The outline of the house remains the same, but the floor plan has been changed around to coordinate better with the new location.

And I can extend my own private deck off the master bedroom right out to the edge of the stream. Much better.

All these changes mean that I had best work up a new floor plan. (You may have to do the same after a close look at your site.) Almost everything that I had before still looks okay, just needs some flip-flopping. But there is just one thing. The big door into the workshop now faces the wind/storm direction. I'll put a small window there instead and move the door around to the north side in place of the window there. That will make it handier for unloading and lugging material inside.

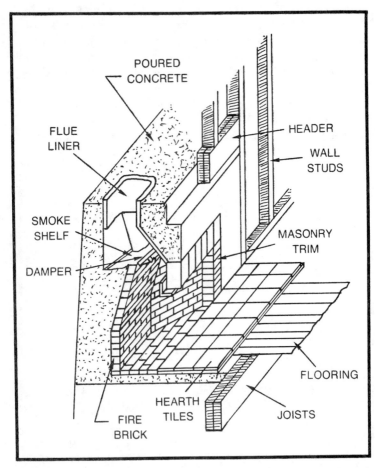

Fig. 4-3. Cutaway of a typical masonry fireplace. There are a good many variations, but the sizes and shapes of the firebox, damper, smoke shelf, and flue are always critical.

ADDING ON MORE DETAILS

A lot of items help to make a house design more pleasing. Here are a few you'll want to consider.

Fireplaces

Today you have an incredibly wide choice in fireplaces. The traditional masonry fireplace (of brick or stone or veneered cement block) with a tile flue is still most popular (Fig. 4-3). They can be scratch-built or worked up around a metal shell. They can be either quite small and unobtrusive or huge and impressive, occupying an entire wall with rock facing, warming ovens, and scrumptious mantels.

The alternative is a prefabricated fireplace, some of which are designed to be built-in, but most of them are freestanding (Fig. 4-4). There is a myriad of sizes, shapes, colors, and styles. They can be set in a corner, stood against a wall, set out from the wall a bit, plunked in the middle of a room, or even suspended from the ceiling at any handy location. Some are equipped with auxiliary blowers to better make use of the heat; others have fully closing glass doors for safety and to prevent heat loss when the fireplace is not operating. Most of them are designed to be used with a double-walled stainless steel chimney pipe, which is not terribly expensive and is lightweight, easy and quick to install, and takes up little room. And it is possible to install one of these units yourself for as little as a couple of hundred bucks, which makes the whole idea attractive.

Storage Areas

And then there is that old problem of storage. Every home has to have storage space for a variety of items, and some families need a great deal more than others. And despite the great need, this is a place where most houses fall woefully short of requirements.

The master bedroom should have ample closet and drawer space. What is ample? Measure the space you now have and estimate how much more you would need for it to be comfortably adequate. A minimum size for a bedroom closet would probably run about 8 feet long by 2 feet deep. A larger closet can have sets of drawers built right in, unless you prefer to use bureaus.

A guest bedroom can get by with a lesser amount of space. Children's rooms are something else. Here you need space to

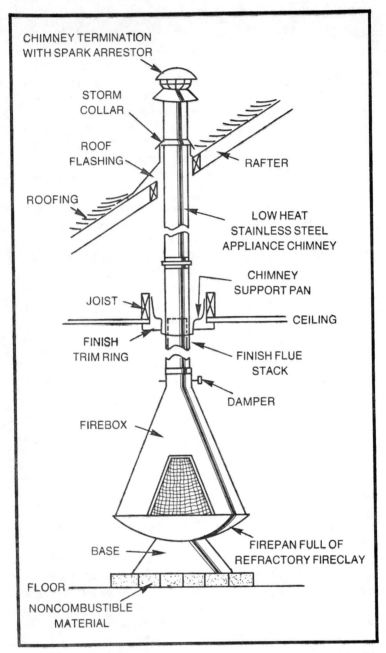

Fig. 4-4. Typical freestanding fireplace and appliance chimney. Chimney arrangement is essentially the same for built-in or wall-mounted prefabricated fireplace units.

hang clothes and lots of room for toys and such. A reasonable amount of closet space is essential, and extra storage can be handled by additional closeting, expanded drawer space, compartments under beds, or shelving along some of the wall space. Figure on enough room for desks or work areas, too, because children equate with school work. And remember that eventually, after the kids have grown up and moved out, those rooms will be converted to other uses, which means that closet and storage space still will be needed, but for different purposes. Make the space as flexible as you can.

Bathroom storage is almost always skimped for some reason, despite the obvious need for space, preferably of the hidden built-in variety. The master, or principal, bath should have provision for storing quantities of toweling, extra supplies, cleaning gear, health aids, sickroom equipment, medicines, lotions, and potions. And maybe room for a hamper and a clothes rack as well. Other full baths need not have as much space but should have at least some to spare. Even half baths should be equipped with medicine cabinet and under-sink storage cabinet at the least.

Heavy outer clothing and foul weather gear is best kept handy to the most consistently used outside entrance, usually the "back" door. A large closet located in a rear hall or as a part of the mudroom is ideal. Make it big enough to contain all such gear for the entire family. An 8-footer is really not too large for a family of four or five. And provide a spot for boots and overshoes, with a special spot for soaked and muddy footwear to dry properly and adequately.

Visitors need a place to hang their wraps too, and the best spot for this is in or near the front entrance. Here you can build in a small closet—3 or 4 feet wide should be ample unless you will be throwing state dinners—perhaps as an integral part of the partition that separates the entryway from the living room. If the closet is big enough, this is also a good place to store your "good" outerwear, as opposed to the everyday garb that hangs in the back hall or mudroom.

Storage space in the dining room is not a necessity because the items that are used there can also be stored in the kitchen. But certainly there is much more convenience in having all of the silver, glassware, china, table linens, and table accessories located near the table. Various pieces of furniture like corner cupboards and glassware cabinets can be used for this purpose, as long as you provide sufficient space for them at the outset. Another arrangement, which can be used alone or in conjunction with furniture, is a built-in section of cabinetry which looks much like those in the kitchen. A large base cabinet can stand on the floor, with wall cabinets hung above. This gives ample storage space for everything, especially if fitted with a few drawers. And the top of the base cabinet can also be used for extra serving space or as a bar.

In the living room, still more storage is needed. You may need shelves for books and racks for magazines, unless you plan to have a separate library. You may need a place for tapes and records, and if you are much of a music lover, you should double the amount of space you think you might need. Shelves for records should be built ruggedly, by the way; records are heavy. And if there is a fireplace, where are you going to stack the wood and kindling and old newspaper? A large built-in woodbox is the answer. And you may have additional specific requirements, like a bar or liquor cabinet, shelving to display your ship models, or an extra closet to hold your slide projector and screen, puzzles and games, card table and poker chips, or whatever.

Lots of folks are collectors of all sorts of things. A large measure of the enjoyment of having a collection is being able to display it to others. And this is an area where many homes fall short. The collector has no place to show his collection, and everything winds up in drawers and cardboard boxes. But with a little planning, suitable arrangements can be made both for present needs and those of the future. In many cases all that is required is some shelving, which can be either of the built-in variety or the type which is simply attached to the walls on vertical rails and adjustable brackets. At the outset, of course, the display requirements can be reduced to just having enough wall space where those shelves can be installed as the need arises. Some collections fare nicely in wall-hung shadow boxes or shallow glassed display cases. Hallways make admirable galleries if the windows and lighting are properly arranged. Many collectibles, such as silver napkin rings, cut crystal glass, or Dresden ware, have their own intrinsic beauty which cries out for permanent covered and lighted display cases, perhaps with mirrored backings. Such can be arranged as built-in shelving modified to meet the needs of the specific collectible and placed in a special area or even a special room.

They can be installed to form a part of the decorative scheme of the living quarters, acting as highlight and accent areas.

Whatever your collections and the particular method of showing them off, work out the details during the house planning and establish them right in the final drawings.

You will also need ample storage space for household items such as linens, bedding, seasonal clothing, and unused toweling. This calls for a linen closet. The best arrangement is a large walk-in kind which really amounts to a tiny room, fitted out with plenty of shelving, drawers, and hooks or racks. Ideally the room should be panelled entirely with thin aromatic cedar planking, left completely unfinished. And since a linen closet is a comparatively inactive storage area as a rule, it can be located almost anywhere in the house that is reasonably convenient.

The kitchen is the room which contains the most storage space, and having too much is a virtual impossibility. Skimping here means disaster over a period of time because a kitchen without ample storage space is a failure. It is unworkable, a constant source of irritation and inconvenience. And unfortunately, a kitchen once built is practically unexpandable (unless expansion allowances have been made in the original design) without a major renovation.

As an absolute minimum, I would suggest that a family of two in a small house should have no less than about 10−12 linear feet of 2-foot deep base cabinets, all with workable counter tops. This does not include the space taken up by stoves, ovens, trashmashers, dishwashers, or other appliances. For a family of four, the minimum linear footage should increase to about 16 feet or more. The wall cabinets above the counters, which are usually only about a foot deep, should rise from approximately 18 inches above the counter all the way to the kitchen ceiling, or to a top shelf height of perhaps 6 1/2 feet maximum. The linear footage here should at least match that of the base cabinets and should preferably be a bit more. And though this may sound like a tremendous amount, it is not, and can be built into even a small kitchen with proper planning.

You should also provide space for all the cleaning materials used in every household. You'll have mops, brooms, vacuum, dustpan, waxer, etc. And there has to be a handy place to put them. The good old broom closet still has its uses and can be tucked into some otherwise marginally useful spot near the work center. Put a lock on it to keep the kids from drinking the floor polish.

How about the possibility of including a pantry in your plans? This was originally a small room or even a large closet which was used for storing kitchen paraphernalia and foodstuffs of all sorts, especially the bulk staple items. Often counter space was provided as well for preliminary food preparation before going into the kitchen. There was a time when no proper large house was complete without a pantry. But along with summer kitchens, borning rooms, and formal parlors, the pantry has all but disappeared.

But this does not mean that the usefulness of a pantry has also disappeared. On the contrary, the idea is staging something of a comeback. The inclusion of pantry counter space is not really necessary, provided that there is enough in the kitchen proper. But floor-to-ceiling cupboards and drawers in a pantry can be ideal for storing seldom-used kitchen equipment, extra glassware and china, and a vast supply of foodstuffs. If you can crack out enough space for a tiny extra room off the kitchen, a pantry is well worth thinking about.

Then there's all that stuff that invariably accumulates in any household. Trunks, suitcases, the old bowling bag, a three-legged poker table, toys that have fallen from favor, the stroller, Christmas decorations, wrappings and boxes—the list is endless. Whether in the attic, down in the basement, or somewhere in the living quarters, you had best make some provision for tucking away all these odds and ends.

Then you will need to allow ample room for the equipment that helps to operate your house. The furnace needs a spot to rest, with room enough around it to negotiate piping or ductwork. If you plan to use individual heating units, these too will take up a certain amount of room. And though they need not be hidden from view, they will take up a considerably larger area then their physical dimensions would indicate, because nothing should be placed near them. And of course there's the hot water tank (a large house may have two) which should be positioned as close as possible to the main points of hot water usage. The freezer needs a spot too.

The workshop speaks for itself. Anyone who uses a shop knows full well how important storage space is. You may need racks for long pieces of lumber, drawers for nuts and bolts and assorted hardware, shelves and racks for tools and equipment.

And even if you don't plan to have a shop as such, in almost every household a certain amount of hardware accumulates, even if only from hanging pictures and fixing the faucets. Some small area somewhere should be designated for this collection.

The same is true for the lawn and garden tools. Almost every household has them, some much more than others. This equipment is best stored outside the house. You could plan for a separate garden shed, add a shed to the greenhouse, add a shed to the back of the house or the garage, or make the garage large enough to house all the gear.

Home offices need file space, bookshelving, and a spot for supplies and papers. Activities areas also need sufficient space to store away all the material and equipment peculiar to the activities that take place there.

And so on and on we go. As you can see, this business of having a place to put things is no insignificant matter. There is nothing worse than a house with a paucity of tuckaway spots. A plethora is better.

Furniture

Once you have figured out how to provide for all your storage needs, you can figure out how your major furnishings are going to fit into the layout. Probably you have a certain amount of furniture and appliances right now, most of which you don't want to part with. And you probably also have certain things in mind that you would like to add in the future. Now is the time to determine whether or not everything will fit into the rooms you have designed.

Go back through your plans and sketch in the major pieces of furniture. If some of the furnishings are to be built-ins, put them in place too. Provide for all the major appliances in the kitchen and laundry. Draw in the cabinets, bars, planters, halfwalls, divider screens, and such. Beds, bureaus, bookcases, dining room tables, buffets, easy chairs, couches, secretaries, desks, highboys, and corner cabinets—anything that has size of any consequence should be put into place in their approximate locations. You may discover that you have some problems, like a grand piano in a 10- by 15-foot living room. Some of the rooms may actually have to be enlarged, or you may want to change their shapes in one way or another. Now is the time, not when the moving man rumbles up with a full van.

Plumbing Fixtures

While you are about it, draw in the plumbing fixtures. Wash basins or lavatory stands, laundry tubs or sinks, water closets, vanities or wash basin counters, bathtubs, showers, saunas and whirlpool baths, the whole works. Again, you may have trouble with the fit. The key thing to remember is to allow plenty of room for the unit itself, plenty of room for the use of the unit, and plenty of room for traffic to get around the unit. But all the elements of furnishings and appliances should create no awkwardness of design and no wasted space. Both are expensive. And by the way, if you have trouble determining the size of some items, consult a Sears catalog. You'll find it to be an encyclopedia of dimensions for many home furnishings and fixtures, perfectly suitable for approximate calculations.

Doors

We have the doorways, and now we need something to close them off. Interior doors usually swing into the rooms which they serve. Closet doors almost always swing outward, away from the closet space. Exterior doors almost invariably swing inward, into the building.

As you ponder your plans, you will doubtless discover that there are some spots where you can simply have doorways—no doors. Make the openings a bit oversize, trim them out in arches or whatever, and just generally pretty them up. In other spots, you may opt for halfwalls, divider screens, or planters instead of the more rigid and formal wall-and-doorway combination. Consider door installation and placement very carefully.

You might also consider the fact that you don't have to have conventional single-swing doors. You can use double-hinged doors which swing both ways. If you want them for effect rather than complete closure, you can use the old saloon-style batwing doors. You can use bifold doors which hinge in longitudinal halves or quarters and take up less space when opened. You can use folding doors which collapse accordian style into a small bundle at one side of the doorway. Or you can use pocket doors, if you plan ahead for them. Pocket doors are solid panels which slide back into a cavity in the wall, disappearing completely. These use up none of the living space and create no obstructions, but do require a wall somewhat thicker than normal.

Stairways

There are two types of stairways. The first is the *conventional* type, a series of steps and risers built into the building as construction proceeds. (For the record, a flight of stairs is unbroken by a landing or any sort of platform; a staircase is bordered by walls or rails and may or may not include landings or platforms.) The simplest form is called a straight run, going directly up from one level to the next. Beyond this, there are a hundred variations, including wide or narrow U's, long or short L's, circulars, or winders. All can be made with combinations of platforms and landings and tread/riser patterns.

Any of these conventional stairways require large chunks of space, and the usual procedure for multiple level access is to place one stairway above the other in order to lessen wasted space. A width of 2 1/2 feet is considered quite narrow, 3 feet acceptable, 3 1/2 feet fine, and 4 feet excellent. If rails or balustrades are attached to the outside ends of the stair treads, then a bit more width should be added. As the stairway becomes more complicated, it takes up even more space and has to be carefully fitted to the basic structure.

The second type of stairway needs a good deal less space but is also much more difficult to navigate. This is the *prefabricated spiral* type, usually made out of steel. The treads may be of steel or wood, possibly carpeted, and often the risers are eliminated. The rise from floor to floor is essentially vertical, so the entire unit takes up little room by comparison to a conventional staircase and gives a much more open and informal feeling. Exact sizes vary, but all that is needed is a trimmed-out hole about 3 feet square in the floor being served. Installation consists of standing the unit into place and bolting it down to the first floor and to the edges of the opening in the second floor. In the case of multiple levels, however, spiral staircases are somewhat difficult to stack.

ELEVATIONS AND PERSPECTIVES

After drawing in plenty of details, my floor plan is finally starting to look workable (Fig. 4-5). But I need to have a better idea of what my house will look like when it's built. I'll need to draw some *elevations* and *perspectives*. Perspectives are drawings that give the illusion of length, depth, and height and are done in proper proportions and in angular perspective,

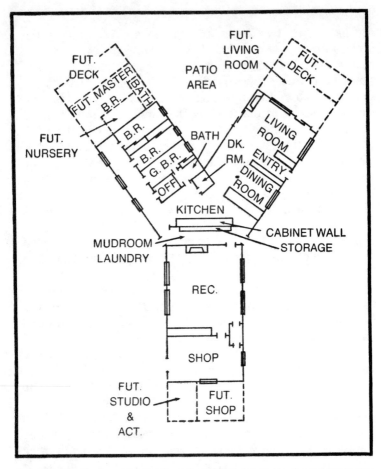

Fig. 4-5. The final rough sketch of the Y house, showing the rearranged interior, door swings, storage, fireplaces, and another expansion scheme.

giving a pictorial presentation. An elevation, however, is a geometrical projection on a vertical plane, showing only one dimension. In house plans, an elevation shows one flat face of the building, including foundation, wall, and roof, usually from grade level on up, as though you were standing directly in front of that face. No other part of the building is shown. Thus, in Fig. 4-6 you see a gable end elevation of a Cape Cod house. Figure 4-7 shows a front side elevation. Usually these elevations are referred to in terms of the compass direction in which they face, such as a north elevation or southeast elevation, or by some convenient descriptive term, like rear

Fig. 4-6. A gable-end elevation sketch of a Cape Cod house.

elevation or entrance elevation. But a perspective of the same house would look something like Fig. 4-8.

In order to make these drawings, you first have to decide what the house will look like with respect to walls and roofs. You have four choices with walls: straight up and down, slanted in, slanted out, or combination roof/wall. The straight are by far the most common.

The choice in roofs is a good deal wider. There are a number of forms which could be used easily on any given structure, plus a few more that are somewhat limited. And

Fig. 4-7. A front elevation sketch of the same Cape Cod house.

Fig. 4-8. A perspective sketch of the same house.

each gives the structure an entirely different look. In Fig. 4-9 you can see that types A through D are adaptable to most structures. Gambrel (E) and mansard (F) roofs are suited to two-story (or more) homes. Note that in these two, as well as the newer version of the A-frame, parts of some walls also become roof. With the geodesic dome type of building it is hard to tell where the walls leave off and the roof begins. Most of the larger geodesics, however, have short vertical wall sections around the bottom called riser walls.

Then there are the various combinations of roofs. There is no law that says that a structure must be under one continuous roof, and in fact, a great many are not. You can use a series of shed roofs, for instance, two or more going in the same or in varying directions. Gable roofs commonly are equipped with gable-roofed or shed-roofed dormers of varying sizes. Two or more gable roofs could be used with varying pitches and be run in different directions. Invert a gable roof and you have a butterfly roof, and the sections of either type need not be of equal length or width. Flat roofs and shed roofs can be used together or in combinations with other types. And of course there are the free-form types which consist of a multitude of compound curves or arches or domes. There is no harm in trying on some of these different roofs for size, just to see how they might look.

For many people, drawing perspectives can be difficult. If this is a problem for you, there is a solution: old cereal boxes and glue. You can while away some of those long winter

Fig. 4-9. Various types of roofs used in residential construction. (A) Flat. (B) Shed. (C) Hip. (D) Gable. (E) Gambrel. (F) Mansard. (G) Tent.

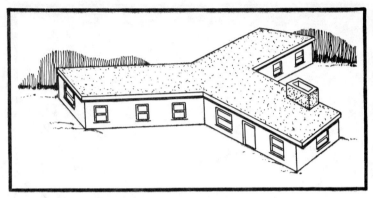

Fig. 4-10. The Y house with flat roofs on all three sections.

evenings by pasting together a cardboard mockup of your proposed home. You can attach various roofs, porches and decks, chimneys, and whatever else suits your fancy, until you come up with something that satisfies you.

For example, there are a number of roof possibilities for the Y house. I could leave the entire roof almost flat, with just sufficient pitch for drainage (Fig. 4-10). Or I could use either a steep-pitched or a shallow-pitched gable roof (Fig. 4-11) or something in between (Fig. 4-12). Or I could put on three shed roofs, starting at the central point and sloping lengthwise down each arm to the ends, highest in the middle and dropping to just above wall height at each end wall. I could use a gable roof on the tail of the Y, with shed roofs on the two arms of the V, sloped to the outside of the V and matching the pitch of the

Fig. 4-11. The Y house with gable roofs on all three sections.

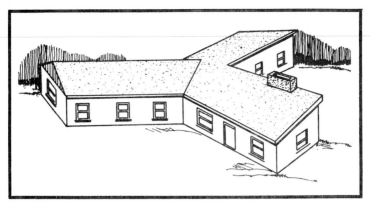

Fig. 4-12. The Y house with gable roof on the tail, and matching-pitch shed roofs on the two arms.

gable section. I might be able to work out a clerestory arrangement (Fig. 4-13). With this setup, one section of roof rises higher than another, and the vertical riser between the two contains windows which can be used for extra light or ventilation.

To draw a complete elevation, you'll probably have to include grade levels. If you plan to build upon a full basement, a partial basement, piers, full footings, or something of that sort, the level of the first floor will be a step or two, or perhaps

Fig. 4-13. A clerestory roof. One shed roof section rises above another.

95

several feet, above the level of the ground. If you build on a concrete slab, the finish floor level will probably be only a few inches above grade level. In either case, however, the grade level can, if desirable, rise above the first floor line. This would be the situation with a house built partly into a hillside, for instance, or where earth is pushed against the outside of a structure to form a sloping bank. Whatever the situation with your building, the elevations should show the details.

Once you determine what your walls, roof lines, and grade levels are to be, at least approximately, then you can run off some elevation sketches as well as some perspectives (or models). Follow your preliminary floor plan layout to position windows and doors, and set the decks and porches or any other architectural features needed. This is a good time to establish some basic sizes, rough heights, and wall positions of the windows too, so that the effects and proportions will look good from the outside.

The Facts of Construction

5

Now comes the question of whether or not this dream house that you have in mind can be built and built for something less than a king's ransom. And whether or not, after it is built, it will stay up there where it belongs and not come tumbling down about your ears some dark and windy night.

Obviously, there are certain physical laws which cannot be controverted and rules of engineering which must be observed. If you try to build a 30-foot-wide roof using 2 × 4s, for instance, not much time will pass before the whole affair becomes a ground-level deck. Walls have to be strong enough to hold up the roof structure; floors need to be sturdy enough to support both you and your furniture; and it's not nice when you can't get the inside temperature over 45° on a snappy winter morning.

You don't have to be an engineer to finish out the design of your home to a reasonable and workable point, nor do you have to be an architect. When the proper time comes, you can find this expertise. When you need something done that is outside your ken or skills, you hire it done. Find a competent architect, one with whom you can really discuss things. If your project is particularly complicated, you might also have an engineer sit in, though this is not usually necessary; a good architect can usually double in brass, and where he cannot, he can hire his own outside expertise. Also you'll need a good contractor. He is the one with the field experience, the practical grasp of construction, the front-line knowledge of what to do and how to do it.

At the moment, though, you don't need all these folks, congenial though they may be. What you do need is some understanding of house construction, how things are put together, what you can do and can't do, what choices you have in materials and equipment, and so on and on. Then you can arrive at some final design conclusions, and *then* you can call in the rest of the boys to polish things up for you and make sure that you haven't gone too far astray.

FOUNDATIONS

To most people, a foundation means the walls of a cellar or a basement or some sort of solid wall that runs underneath the perimeter of a house. A house doesn't necessarily need this. A foundation is actually the base upon which a structure rests, and that need not be basement walls or, for that matter, any kind of walls.

Let's look back in time a bit. Oftentimes in the early days around New England, a rough-cut beam was laid upon a few stones set upon the ground, and the walls went up from there. The foundation consisted of those few stones which quickly heaved about from frost and sank into the soil, which in turn became the foundation. The home builder in a hurry didn't even bother with the few stones. He just notched a furrow into the ground and dropped a log into the groove, then piled up some more logs until he was sufficiently sheltered. Often as not the floor consisted of pounded earth; only the better cabins boasted wooden floors.

This method proved unsatisfactory from several standpoints, however, so some improvements were made over the years. The homeowner dug a shallow trench, laid up a course of fieldstone dry (without mortar) and laid his sill beams upon them. Better, but still problematical. In homes where a basement was needed, the stone work started at the basement floor level and was carried up to slightly above ground level, and the sills laid upon the resulting walls. A darn tough way to build, and by today's standards, neither efficient nor effective.

Further modifications were made, using granite blocks, mortared fieldstone, brick. Finally came the poured concrete walls or the built-up concrete block walls.

A stone foundation laid dry has almost no lateral or sideways strength, and weight pushing down upon it from above tends to separate the stones, squirt them out in one direction or another. Stones laid in mortar are not quite so bad, but nearly so. Stones are difficult or impossible to get in many areas, are difficult to work, create problems in haulage and handling, and competent stone masons are expensive.

A built-up masonry wall of premanufactured materials, such as slump block, concrete block, cinder block, and various types of brick also has little lateral strength. Because of the flat planes involved between the masonry units, they will take a great deal of weight pressing straight down, but they are expensive to build, must be thoroughly waterproofed, and are always damp because they absorb moisture from the air and surrounding soil. The tolerance for this type of wall or foundation to settling, vibration, or shifting is low because the material has so little elasticity. Cracks are bound to appear.

How about that panacea for all ills, poured concrete? This has more advantages than built-up masonry, to be sure. It will stand more straight-down push, has greater lateral strength, is seamless and so a bit more elastic and a bit less susceptible to water leakage.

Now anytime you deal with foundations, you must consider weight. Your house has a certain mass, and that mass must rest on the earth. The earth will only hold so much weight without subsiding, sinking down. The amount of subsidence depends upon the surface area on which the weight rests. That's easy to prove to yourself. If you set the point of a knitting needle on the ground and rest a 10-pound weight carefully on top, the needle will sink into the ground a short way. But rest a chunk of 2×4 endwise on the ground and place the same weight on top, and nothing much will happen. The ground has a certain compressibility, a certain resistance to weight. With the 2×4, the ground loading (the weight on the ground) is roughly 1.9 pounds per square inch, the equivalent of about 275 pounds per square foot, and the ground usually can withstand that pressure. But the needle presents a much smaller surface to the ground. If we say the point of the needle is 0.1 square inches in diameter, then the ground loading with that same 10-pound weight is 100 pounds per square inch, or the equivalent of 14,400 pounds per square foot. That is more than almost any kind of ground can handle.

Spreading a given amount of weight out over a larger area will prevent it from sinking. That's called flotation and is the principle behind using fat tires on dune buggies and snowshoes for winter foot travel.

Now your house will have a certain weight, which you will have to keep from sinking into the ground. Let's suppose, however, that you will use standard frame construction. In any case we are actually concerned with two kinds of weight, live weight and dead weight. Dead weight is the term used for the mass of the house itself; live weight is the variable loading of the occupants of the house and their belongings.

With a lot of calculations and a set of weight tables, you can figure out just what any given house will weigh, simply by adding up all its parts. But I'll give you a good working figure of 100 pounds per square foot of floor area for dead weight. A common live load figure used in the design of single-family residences is 40 pounds per square foot, so that brings us up to 140 pounds for every square foot of floor space in the building. The roof also must be figured in. The dead load is the weight of the roof itself, and the live load is the snow and wind load, which can vary from locale to locale. A figure of 100 pounds per square foot total would not be amiss here and will cover nearly every likelihood for a residence.

For example, I'll take a 1500-square-foot house with an 1800-square-foot roof. That gives us 210,000 pounds for the first part of the formula (1500 × 140), and 180,000 pounds for the second part (1800 × 100), for a total of 390,000 pounds.

All that weight must rest on the ground. How much weight will the ground hold, per square foot? Obviously this varies, too, not only from locale to locale, but often between spots only a few feet apart, or even only a few inches. You can arrange to have tests made on your own building site to determine exactly what the loading capabilities are. You can also use a rule-of-thumb figure which is good in almost all areas, except those with special problems such as marsh, swamp, deep sand, or solid rock. If you keep the loading under 1000 pounds per square foot of ground, you'll be safe. Less is better, of course, and if you are uncertain about conditions, have some tests made.

If the 1500-square-foot house is 30 feet wide and 50 feet long, then there will be 160 linear feet along the outside edge which will require 160 running feet of foundation wall.

Assuming that there are no other supports under the house and the entire weight will rest upon the foundation wall, the ground loading will be 390,000 pounds divided by 160 feet equals about 2438 pounds per running foot. Taking the figures a bit further, the weight of the foundation wall, if it is a foot thick and 8 feet high for a full basement, will itself contribute another 1120 pounds per square foot (the weight of each section of wall a foot wide by a foot thick by 8 feet high) to the ground loading for a total of 3558 pounds per square foot. If the wall is 8 inches thick, a common dimension, ground loading without a footer pad is upwards of 4400 pounds per square foot. Not so good.

Suppose now that you take that same concrete that was poured into the foundation walls and spread it out instead into a big pad 50 feet long and 30 feet wide. Under the edges of the pad is a small footer (Fig. 5-1). Insulation around and under the pad helps keep you toasty, and within the concrete is suitable reinforcing material. This pad is a monolith and distributes weight relatively evenly. Now we have a 1500-square-foot foundation lying flat on the ground. The ground loading has diminished radically to an average of 260 pounds per square foot. If you include the weight of the pad, the figure would go up to only 400 pounds per square foot (even if the pad were a foot thick). Actually, the loading would be higher in some places and lower in others, depending upon how the house was built upon the pad, but the point is the remarkable reduction in ground loading. You have provided flotation, and spread the load around.

Of course, a full pad for a foundation is not the only way to go. If the ground is solid, there's a better and much less expensive method. Use small pads, properly spaced at strategic locations, with posts, or piers, resting upon them; the house can rest on top of the posts. At the 1000-pound rating, a 4 by 4 foot concrete pad would support about 14,000 pounds since the weight of the concrete pad itself would contribute only 140 pounds per square foot if the pad were a foot thick. That means that the 390,000 pound house could be supported by 28 pads. This would use some 14 cubic yards of concrete instead of possibly four times that amount for a full pad or for full foundations. In addition, there is practically no digging, no site disruption, no loads of gravel to be hauled in, no spoil dirt to be hauled off, no concrete finishing.

For the sake of comparison, let me explain two ways you can provide a foundation that employs no masonry or concrete

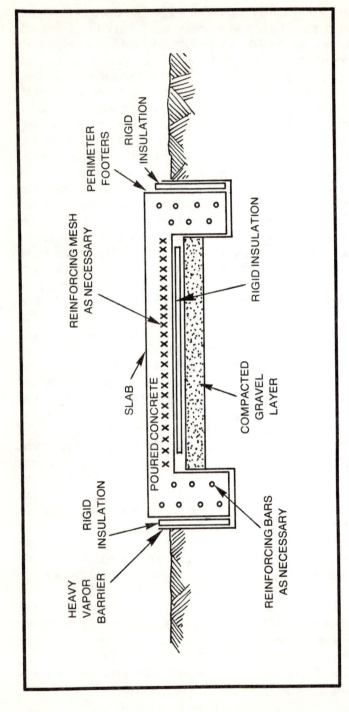

Fig. 5-1. A poured concrete pad provides footers, foundations, and first floor level all in one shot.

at all; it can readily be constructed by one man. Both ways are perfectly valid and produce fine results; both methods are relatively inexpensive.

The first method uses a system of plywood and wood stock (pressure-treated, containing preservatives). The system has been under test for many years and has proven itself. In practice, a special wood footing sill is placed on a narrow, level gravel base at a suitable depth below grade to provide either a full cellar or a crawl space. Then walls are framed in the usual method but using treated 2 × 6s and covered with treated plywood sheets. After the exterior seams are caulked,

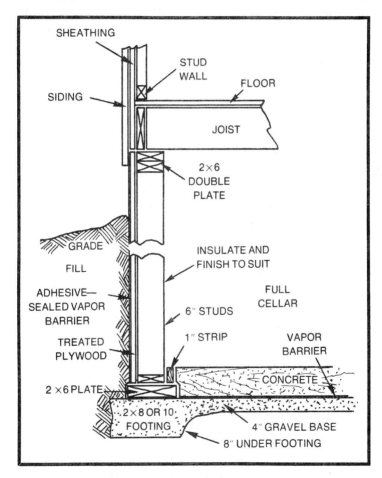

Fig. 5-2. Typical all-weather wood foundation. This method is approved by the FHA and is also now being included in many building codes.

the entire outside below-grade area is covered with a glued-on polyethylene sheet (Fig. 5-2). Then if a floor is needed, either a concrete floor can be poured in the usual manner, or additional treated joists can be secured to the wall frames and a wooden floor laid. Backfill is run into place against the foundation in the usual manner.

The second method is ideal for the do-it-yourselfer. All you need is a post-hole digger, a post-hole bar, and a whole flock of posts. Once the outline of the house and the exact position of each post is marked out, just dig a hole, plant the post and tamp it in securely, remove the little bit of extra dirt, and trim the top of the post to the desired height (Fig. 5-3). Then the first flooring joists can be nailed directly to the posts, and the framing work is begun with no fuss and bother and doggone little expense. And the ease with which the house begins to take shape is amazing. The posts, of course, should be treated with preservative, and the size will vary with the weight of the house relative to the minimum number of posts you want to put in, or the maximum number you might want to provide added support so that you can keep the size of the floor joists down.

This system works beautifully. The reason it does is technically complicated, and I won't attempt to explain it fully. But these posts are actually pilings. The post does not depend upon just the bottom surface area for its holding power, but upon all the sides of the post as well. You may know what a job it is to yank up a stake or a fence post; it takes an incredible amount of force just to pull a small one straight up from the ground, and the deeper it is, the more force is required. The same amount of force, or perhaps a bit more, is required to push it down. This is why docks, piers, bridges, and large buildings are stuck up on posts (pilings).

There are a good many distinct advantages to post foundations besides the complete ease of installation. For instance, you don't have to worry about drainage or ground water or snow piling up. Nothing can touch the house and nothing will bother the posts at all, if they are treated or are naturally impervious. You can quickly adjust the posts to the contour of the land by setting them to rough height and then lopping them all off accurately with a saw after they are in. No site leveling is needed, and you can handily build with this method even on a steep slope, simply by adding some cross bracing as the posts become tall. Virtually no soil has to be

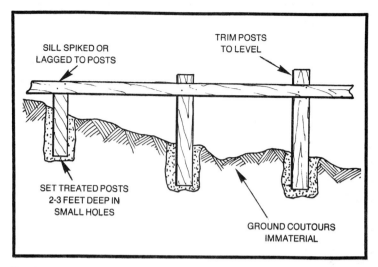

Fig. 5-3. Heavy treated posts set in the ground make a fine foundation. Site disruption is minimal, contours (within reason) make no difference, expense is low, and the installation simple.

removed from the site, and no gravel or other material brought in, save for the posts themselves. Just scrape away the topsoil and dig a small hole. When the building is complete, you can either leave the space between ground level and the house floor wide open, which won't hurt a thing, or you can close it in with latticework or solid skirting. In the latter case, be sure to provide plenty of ventilation openings to keep the moisture down.

A slight variation on this system is the use of poured concrete piers instead of wooden posts. Builders sometimes use this method because it is fast, inexpensive, and efficient for them. They drill the holes with a big power auger, slip in a length of tubular form made from specially treated cardboard, add reinforcing bars if necessary, pour the forms full of concrete, and bed in some anchor bolts to hold the sill plates of the structure.

FLOORS

Time to move up a notch and consider the floors. Floors have to be rugged for two reasons. First, this is where a lot of dead weight and a lot of live load collects. Fifty folks all milling around at a standup cocktail party, with half of them right next to the piano, really puts a strain on things. And

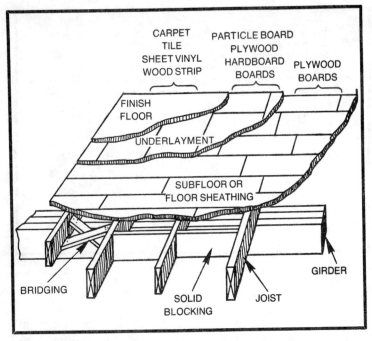

Fig. 5-4. The conventional floor is put together like this. Sometimes there are only two layers of flooring, other times as many as four or five.

second, most of us just don't like floors that feel like a trampoline as we cross.

A conventional floor structure consists of three and sometimes four parts. First, there is a series of heavy planks, the joists, set on edge (Fig. 5-4). On top of them, boards or sheets of plywood are nailed down to form a subfloor, or floor sheathing. If the finish floor is to be tile or vinyl floorcovering or carpeting, often another layer is added in the form of underlayment. This underlayment usually consists of sheets of particle board (a mixture of compressed sawdust and bonding materials) or hardboard (a thinner material made from compressed fibers and bonding agents).

Of course there are other ways to build a floor, less expensive ways. But so you can understand what is involved here, let's first look at some of the characteristics of a piece of wood.

First consider strength. A piece of wood of a certain size has a certain strength. I'll say that a board 2 inches wide and 1 inch thick and a foot long has a strength of X. Another piece of the same material, same width and same length but 2 inches thick, has a strength of not 2X, but 4X (Fig. 5-5). Double the thickness, and it will take four times as much force, or weight. All things equal, the strength of a board increases as the square of its thickness. What this means is that if an 8-foot 2 × 4 is strong enough to hold up 100 pounds, then an 8-foot 2 × 8 of the same material will hold up 400 pounds. This assumes no physical defects in the material like notches, holes, or big knots, by the way. Anything of this sort will weaken the material.

Here's another important point: A beam's ability to carry a load decreases as its span (or length) increases, and in direct proportion. That means that if I take my 2 × 8 which will hold up 400 pounds and extend it to 16 feet, it will only hold up 200 pounds. At 32 feet, the capability diminishes to 100 pounds. The longer the span, then, the thicker the beam must be to hold up a given load.

Next consider wood stiffness. Stiffness is the resistance to bending when a force is applied. The stiffer the material, the less it will bend under a given pressure. Let's say that the 2-inch-thick by 2-inch-wide by 1-foot-long board, supported at each end, will bend a little in the middle when you place a weight on top of it (Fig. 5-6). The amount of bend (or deflection) is X inches. Now substitute a 1-inch-thick board and pile the same weight on top. The piece will not bend down

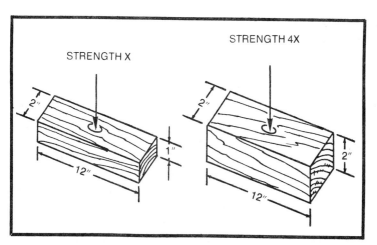

Fig. 5-5. Double the thickness of a piece of stock of a given width, length, and material, and you quadruple the strength.

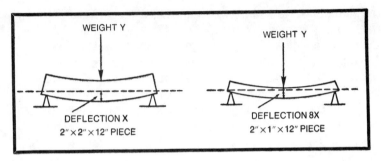

Fig. 5-6. The thinner a given piece of stock, the less the stiffness and the greater the deflection.

twice as far, nor four times as far, but eight times as far, or 8X. The deflection is dependent upon the cube of the thickness of the material. So if my 8-foot 2 × 8 deflects a half inch under its 400-pound weight, then an 8-foot 2 × 4 of the same material will bend down a full 4 inches, or 8 times as much.

Now move the end supports and place a piece upon them 2 inches thick by 2 inches wide by 4 feet long. The deflection will not be X as it was with the 1-foot piece, nor 4X because this piece is four times as long, but 64X. The deflection is also dependent upon the cube of the length of the piece. If the piece were only 2 feet long, the deflection would be only 8X, however. So if I double the length of my 8-foot 2 × 8 with its 400-pound load and 1/2-inch deflection, that deflection will become 4 inches, the same as the 2 × 4 of half the length.

When you boil all this down, here's what is left in the bottom of the pot:

- Double the thickness of a given material, and you quadruple the strength. Triple it, and it becomes nine times as strong.
- Double the thickness of a given material, and it will retain equal strength over a quadruple span.
- Quadruple the span for a given thickness and material, and the strength is quartered; double the span and the strength is halved.
- Double the thickness of a given material, and you multiply its stiffness by eight.
- Double the thickness of a given material, and it will retain an equal stiffness over double the span.
- Double the span for a given piece of material, and the stiffness will be halved.

Let's get back to the floor and put some of this information to work.

A subfloor is often put down first to give the workmen something to work on; it will later be covered up by more flooring. Underlayment is frequently put down because the subfloor isn't thick enough to begin with to get everything up to the proper level and thickness, and because different thicknesses of finish flooring have to be matched up to the same level to present an even surface. And furthermore, the subfloor, usually 1/2- or 5/8-inch plywood, is not stiff enough to avoid bounce.

But you can do away with the whole silly business. Cover the joists with a single layer of tongue and groove 2 × 4 or 2 × 6 decking planks. Protect the upper surface with heavy plastic sheet film or building paper, then give it all a light sanding, and finish the whole thing with a floor oil. Makes a dandy floor, sturdy as a gymnasium, for less cost.

A single layer of flooring, whether you later install carpeting or tile on top or not, makes a better floor for less work because of the stiffness gained. Or to turn that around, when you install multiple layers of flooring, you lose both strength and stiffness compared to a single thickness of the same material.

If a piece of material, say a 1-inch-thick plank used for flooring, has a stiffness factor of 100%, then a plank 2 inches thick will have a stiffness of 800%, as explained earlier. When you start going backward, strange things happen. A plank of half thickness will have a stiffness of only 12.5% of the original, and a 3/4-inch board will be only a bit over 42% as stiff. So two 3/4-inch boards, even though they have a thickness 1/2-inch greater than the 1-inch board with 100% stiffness factor, will still have only a bit more than 84% of the stiffness of the single board. And note that the stiffness of the two 3/4-inch boards is only about 25 percent as great as a single 1 1/2-inch board of the same material.

The loss in strength is not so great. But neither, in most residential cases, is it as significant as the loss of stiffness. If the 1-inch board again represents 100%, then doubling the thickness will jack up the strength to 400%. Cutting the thickness in half will give you only 25% of the original strength, and a 3/4-inch board will be just over 56 percent as strong. So here, two 3/4-inch boards will be about 112 % as strong as the

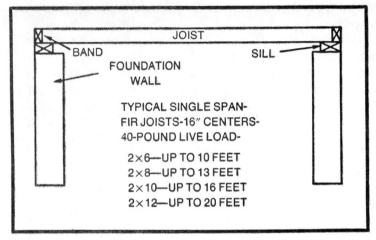

Fig. 5-7. The span of any given joist is dependent upon the live and dead loads it will carry, the thickness and depth of the member, the material it is made from, and its quality.

100% 1-inch board. But again, the strength of those two boards is only half as great as a 1 1/2-inch piece of the same material.

It makes sense, then, to go with *one* layer of flooring over the joists, if you can reconcile that with your general plans. I mentioned that you could use 2-inch-thick decking material, which is really only a tad over 1 1/2 inches thick after it goes through the mill and gets planed down. But there is another type of planking called five-quarter, designated 5/4, which is worth looking into as well. Several kinds of wood are available, the thickness is actually a full 1 1/4 inches, and this makes fine flooring.

Direct your attention also to the underpinnings of the floor, usually called the floor frame or floor structure. Here we have the joists that support the flooring and also the supports that hold up the joists. Conventional underpinning is shown in Fig. 5-7: A series of joists span from one foundation wall to the other. But many times additional support must be given to the joists in order to bolster up the stiffness and strength (Fig. 5-8).

With proper support, you can diminish the size of the joists and still have equally strong and stiff floors. By using a somewhat different method than the conventional one, you may be able to diminish the sizes a bit more and end up with an even stronger, stiffer floor which is easier to install.

Let me sidetrack for a moment to explain a point which is of considerable importance. Wood is the most versatile building material we have for residences; it is most unlikely to be superseded to any great extent in our lifetimes. When you buy wood planking or wood framing material, you buy it by the board foot. A board foot is a section 1 foot long, 1 foot wide, and 1 inch thick. In actuality, the board foot you buy today is something less than that, thanks to the lumber industry which has pared a 2 × 12 down to about 1 5/8 by 11 3/8 inches, with similar reductions in all other stock.

So, every time you figure out a way to save a board foot of lumber, especially if you can do so without sacrificing any strength or stiffness or looks, you are doing several things to the good. You are saving yourself some money. You are conserving, even though in a small way, a natural resource. In some cases you may also be saving some labor costs.

Remember too that planks and dimension stock come in length multiples of 2 feet, usually starting at the smallest size of 8 feet and going up from there. For pieces beyond about 20 feet, you begin to pay an increasing premium as the length increases. So if you design a house section which will require a series of joists that are 12 feet 4 inches long, you'll have to buy 14-footers and cut off 1 foot 8 inches from each one. It's better to change your design to require an even 12 feet, or expand to an even 14 feet.

Now back to the floor frame. Let's assume you have a span of 18 feet that you can bridge adequately with 2 × 12s, and you

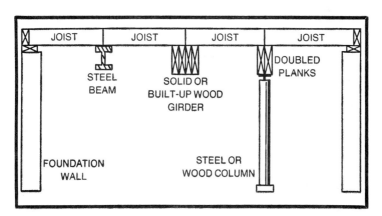

Fig. 5-8. By using various methods of supporting the joists, the members can be made smaller, lighter, and less expensive.

will need 10 of them on 16-inch centers. Each plank contains 36 board feet of wood, so you will have to buy 360 board feet of material. At a price of $250 per 1000 board feet, this will cost you $90. This also happens to be almost the maximum span for this timber, so you know ahead of time that there will be some bounce to it.

Suppose you provide a support midway under each joist, a good solid one, thereby halving the span. Now you can reduce the size of the joists to 2×6, achieve the same purpose, gain some strength, and cut the cost in half. Then you must add back the cost of the supports, but you'll still come out ahead.

Another way to build a floor is to use a grid system. This works particularly well for a building which has a crawl space underneath. With proper supports, you can even use 2×4s all the way through and come up with a floor strong enough to hold up a herd of wild mustangs. This system uses a series of small joists laid in the usual fashion between the sills or main floor beams. On top of this, lay down a layer of Celotex sheets, or some similar composition building board (Fig. 5-9). Be careful where you step, or you'll whistle right on through. On top of this, nail down another series of crossjoists running in the opposite direction. Now you have a gridwork of tightly joined 16-inch squares, which provides an amazingly strong and stiff structure. At this point you can run pipes and wires in the spaces (being careful to remember that pipes have a tendency to freeze), and stuff in some insulation. There should also be, somewhere in there, a skin of aluminum foil, shiny side up or shiny on both sides. An additional vapor barrier of polyethylene film wouldn't hurt, either. When you are done messing around with the innards of the structure, you can lay single-layer flooring on top of the grid, which can then serve as the finish floor or a base for tile or carpeting or what-have-you.

If you feel uncomfortable with all these facts and figures, you can always ask your architect. An architect/engineer can consult his charts and tables, slip his slide rule a few times, and tell you plenty about what beams you can use where and what combinations of stock you can use to achieve X stiffness with Y strength on Z supports under Q loading, and all those good things.

WALLS

In conventional wall framing, one starts with a single 2×4 laid flat on the floor. This 2×4 is called a sole plate. A

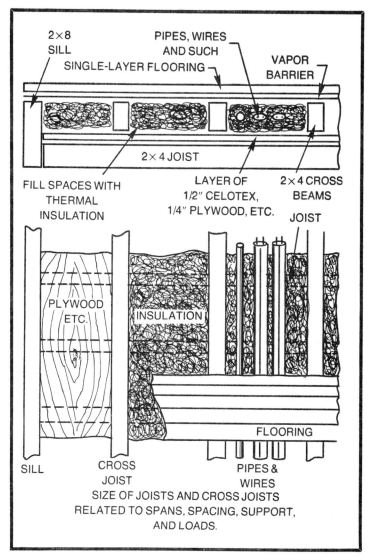

Fig. 5-9. One way to build a grid floor system using single-layer flooring.

series of 2 × 4 studs, generally on 16-inch centers, are then stood up on the sole plate, and a top plate consisting of two more 2 × 4s (one atop the other and both laid flat) are then nailed to the tops of the studs. Where door and window openings are necessary, a header is put in at the top of each opening, trimmer studs run down the sides, and cripple studs

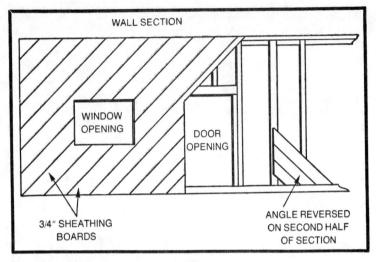

Fig. 5-10. Diagonal plank sheathing on walls was commonly used before the advent of plywood and sheathing sheets. It still is an excellent method, resulting in a strong shell and sometimes lower costs.

are put in above the headers if there is room. With windows, cripples are put in both above and below, under the window sill. Then blocking is nailed in between the studs, and a sheathing is applied. Usually the whole business is assembled flat on the floor and then raised into place. This is called platform framing and is a simple and convenient method. There is another system called balloon framing which is different in detail but quite similar in result. It has some advantages, but also a few drawbacks, and is not much used nowadays.

But there is nothing graven in stone that says a wall must be built just this way. In some designs it may be possible to eliminate the top layer of the top plate, using just a single 2 × 4. Save a little bit that way. And most houses now are sheathed with plywood or with building board (a soft, fibrous composition board, light in weight, not very strong, 1/2-inch thick), or a combination of the two. Because the building board is so flimsy, most builders will nail sheets of plywood at the corners of the structure for added strength and rigidity and use the building board in between.

You don't have to use either one. You can sheath with plain old boards, as was done successfully for a good many years. If you are working by yourself the task is much easier than juggling big sheets. The planks can be laid straight across the studs, or in a diagonal fashion for greater strength and rigidity, as shown in Fig. 5-10. You may find that the total cost is less for 3/4-inch planking than for the usual 1/2- or 5/8-inch plywood, and you might get a better job to boot. Plywood is not necessarily stronger than regular planks. It can be stronger in some ways, under some conditions, for some uses. But that does not make plywood the lead choice under all circumstances.

There are some variants of the old post-and-beam, or barn, walls that will work nicely too. Instead of studs and plates, you set up wide-spaced posts (Fig. 5-11). The exact size and spacing is determined by the weight of roofing that must be held up, as is the size of the top beam which connects them. Then, you can run crossbeams laterally across the posts and attach vertical boards on the outside to form the outside finish wall. Or you can apply sheets of material.

You can modify the above design by using two layers of boards. One layer of boards is applied vertically, nailed top and bottom, with no crossbeams used. At window openings, the boards just dangle there, which seems pretty flimsy. But the next move helps that situation. Nail, or screw, horizontal narrow strips of wood about every 2 feet. The wood can be scraps long enough to reach from post to post (Fig. 5-12). At the same time, frame around all the window or other openings in the same way. When this is complete, nail up another set of vertical battens, spaced so that vertical exterior boards can then be nailed to the battens. Insulative material can be sandwiched in between layers as you go, as necessary. This can be reflective foil in one or two separated layers, wool batt material, or rigid planks of expanded polystyrene foam plastic. And if you prefer to have the exterior lines of the planks going sideways rather than up-and-down, apply the inside layer horizontally, the furring strips vertically, then run a series of properly spaced nailing strips horizontally, and nail siding to the strips.

The mention of insulation brings to mind a recent development in the construction of wall sections. The method of construction is the standard platform framing discussed earlier. The difference is in the size of the framing material, or "sticks." Instead of the usual 2 × 4s on 16-inch centers, use 2 × 6 stock on 24-inch centers. And of course all the other

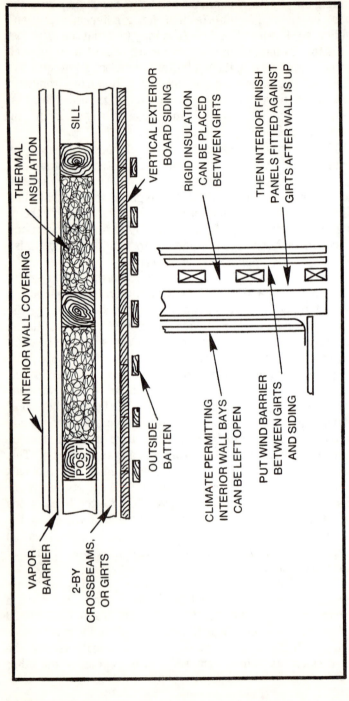

THERMAL INSULATION

SILL

INTERIOR WALL COVERING

VERTICAL EXTERIOR BOARD SIDING

RIGID INSULATION CAN BE PLACED BETWEEN GIRTS

THEN INTERIOR FINISH PANELS FITTED AGAINST GIRTS AFTER WALL IS UP

POST

VAPOR BARRIER

2-BY CROSSBEAMS, OR GIRTS

OUTSIDE BATTEN

CLIMATE PERMITTING INTERIOR WALL BAYS CAN BE LEFT OPEN

PUT WIND BARRIER BETWEEN GIRTS AND SIDING

Fig. 5-11. Post-and-beam wall using girts and a board-and-batten exterior.

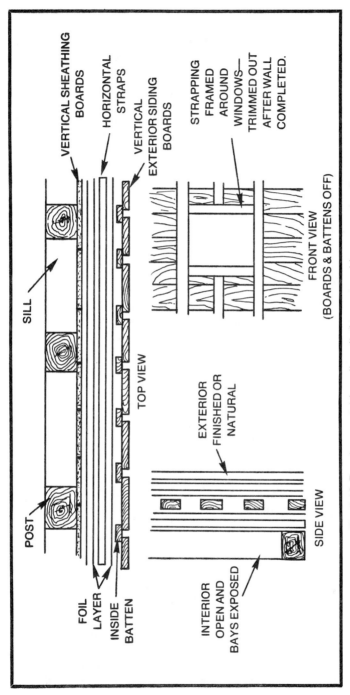

Fig. 5-12. Two-layer outside wall applied to post-and-beam construction.

VERTICAL SHEATHING BOARDS

HORIZONTAL STRAPS

VERTICAL EXTERIOR SIDING BOARDS

STRAPPING FRAMED AROUND WINDOWS—TRIMMED OUT AFTER WALL COMPLETED.

POST

SILL

FOIL LAYER

INSIDE BATTEN

TOP VIEW

FRONT VIEW (BOARDS & BATTENS OFF)

EXTERIOR FINISHED OR NATURAL

INTERIOR OPEN AND BAYS EXPOSED

SIDE VIEW

117

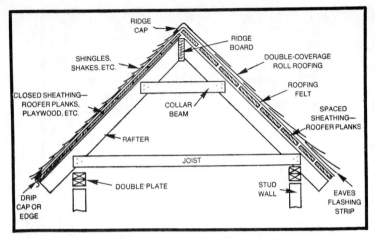

Fig. 5-13. Conventional roof construction.

elements of the wall are also 2 × 6. The relatively deep wall opening is then filled with wool batt insulation with an R-19 rating. This is one of the steps used in building a house which is amazingly energy conserving.

As you can see, you have some flexibility of choice with exterior wall sections, just as you do with floors and foundations. By picking and choosing, developing some of your own ideas and using your imagination, relating the results to your overall design and your environment (including the weather and climate in your area), you may be able to find a combination that suits you and your pocket book to a T.

ROOFS

Onward and upward. As you might expect by now, you have some choices when it comes to roofing, too. Roofs today are almost invariably built in one of two ways.

The first is the most conventional and uses a series of rafters—common, hip, jack, valley—which are nailed together. If the roof is flat or shed-type, the common rafters merely lie across the wall top plates (Fig. 5-13). As with floors, beyond a certain span, additional support of one sort or another has to be provided.

The second method uses roof trusses (Fig. 5-14) which are prebuilt to the necessary specifications and then installed as units. The upper part of the truss serves as roof rafter; the bottom chord becomes the ceiling joist; and all the mess in the middle is bracing to strengthen and stiffen the whole. Depending upon the particular job in question, trusses often are advantageous from the standpoint of possible span length, overall strength, or cost.

With either method, the next step consists of a layer of sheathing. Commercial builders usually use sheets of plywood because this saves them time. Planks, however, work just as well, may be less expensive, and can provide equal or greater strength or rigidity, depending upon specific circumstances. In some of the sunnier climes, narrow strips are used, with a wide space between each one. Then the finish roofing is applied over this roof deck. Tried and true methods, both of them, and they work.

But there are other methods too. The first thing to consider is what a roof has to do, and consequently what it must be. Primarily, it keeps out the weather. It supports practically no load, by comparison with a floor. It has to hold up itself. In some climates it must also hold up a certain amount of snow. It must resist a certain amount of wind pressure.

So strength is a consideration, but not great strength. Design plus local conditions might indicate a loading of 100

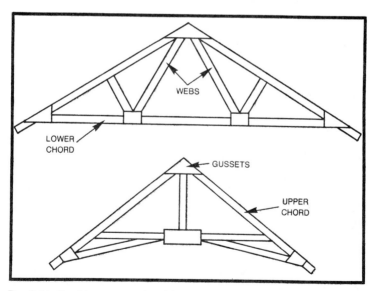

Fig. 5-14. Roof trusses are in widespread use today and have numerous advantages. They afford greater interior design flexibility with the elimination of load-bearing interior partitions, and they are strong and easily erected.

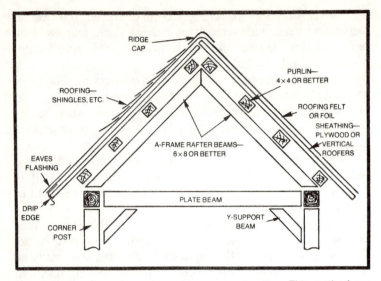

Fig. 5-15. Purlin roof in post-and-beam construction. This method can also be adapted to conventional platform-framed buildings.

pounds per square foot (fairly high), or perhaps only 50 pounds or even less. The closer you can calculate and the lower you can reasonably get this loading figure, the better off you will be. The object is to use as little material, both in pounds and in board (or square) feet of structural material, as will do the job. And note that stiffness is of relatively little consequence. Who cares if the roof is a little bouncy? There won't be anybody up there to worry about it, and there won't be any adverse effects.

One method which accomplishes this purpose quite nicely is another form of post-and-beam construction. You can, for instance, position a series of heavy rafters at wide intervals and cover them with a special roof decking material which comes in sheets, serves as its own insulation, and effectively locks together to form a solid deck. The finish roofing is then applied over this, and the underneath can be left as is or covered with a decorative skin.

You might also position a series of even heavier rafter beams even further apart, as was done in the old A-frame style of building (Fig. 5-15). Then lay a series of crossbeams horizontally across them, again with fairly wide spacing. These crossbeams can be much smaller than the main rafters, incidentally. The roofing planks, or roofers, are then nailed vertically onto the crossbeams, so that they contribute the greater part of their stiffness where it is most needed, over the longer spans. Pound for pound, this will do as well or better than any other system. Sheets can be substituted for planks if desirable, and insulation placed as necessary, with the underside finished or left open.

Now let's talk about pitch. Pitch is the degree of slope of a roof and is designated in terms of the number of inches the roof rises over the number of feet of span that the roof covers. Thus, if you have a shed roof which will cover a structure 20 feet wide, the run of the roof (the horizontal travel) is also 20 feet; the eaves, or roof overhang, is not included. If the roof were pitched to a peak in the middle, then the run would be 10 feet for each side (Fig. 5-16). Now, in the case of the shed roof, if the roof rises a total of 20 inches from one side to the other, over the run of 20 feet, then the pitch of the roof is 1 inch to the foot, or a one-in-one pitch. But if a gable roof rises a total of 20 inches to the peak, it does so from each side over a run of only 10 feet. This would be a pitch of 2 inches to the foot, or a two-in-one pitch (Fig. 5-17). If a roof rises 40 inches over a run of 10 feet, the pitch is four-in-one.

This business of pitch is an important one from several major standpoints. In the first place, pitch determines how well water will drain off the roof, instead of into the house. A flat roof is just a disaster waiting for a good opportunity to happen; drainage is no better than in a parking lot. Puddles occur, installed drains plug or freeze up, ice accumulates. On

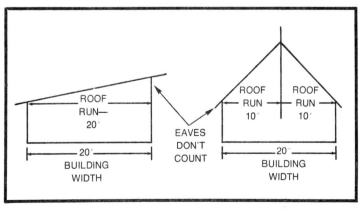

Fig. 5-16. The pitch of a roof depends on how many inches the roof rises per foot of its horizontal span.

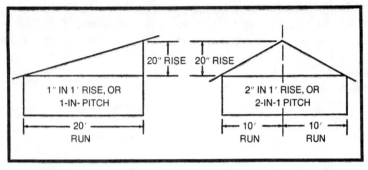

Fig. 5-17. A roof with 1-in-1 pitch and a roof with 2-in-1 pitch.

top of that, a flat roof has to be structurally stronger and better braced than a pitched roof. The weather surface has to be built just so, and out of the best materials, and with the best workmanship, in order to hold up for more than a few short years under the constant pounding it gets from sun, storms, expansion and contraction, and folks tromping around on it. Despite the advertisements, flat roofs are just a big pain in the neck, and I can say that from considerable dreary experience. And if you've never shoveled four feet of snow off a flat roof to keep it from caving in, you haven't lived. Scratch the flat roof off your list.

Just a little bit of pitch is a big help because then gravity goes to work and the water runs off. As you work upward from one-in-one to about three-in-one, the drainage action gets a little better, but not remarkably so. Over four-in-one the water fairly flies off, and when you get up around ten- or twelve-in-one, the snow will zip off as well, depending upon the snow's composition and the weather. For this reason, the rapidity and effectiveness of drainage, certain roof covering materials must be used only above certain minimum pitches (Fig. 5-18).

Zero-pitch roofs are usually covered with a so-called built-up roof skin, which consists of multiple layers of roofing felt (not cloth, but a heavy fibrous paper impregnated with asphalt, which comes in rolls and is designated by its weight per 100 square feet, e.g., 15-pound felt, 30-pound felt, etc.) all stuck together with a coating of hot tar between each layer of felt. On top of this goes another thick coat of hot tar and a topping of fine gravel. This same construction is frequently used in pitches of up to one-in-one or a bit more. Above this pitch, and up to about one-in-four, you can use double coverage roll roofing. This is heavy asphalt-impregnated material, usually with a mineral surface (tiny chips of stone) on the weather side. It is applied in horizontal layers with the top layer overlapping the bottom one by nearly half, with a tar coating along the edge of the overlap. Actually, this same type can be used with all steeper pitches, too, though it is not necessary.

From a four-in-one pitch on up, single coverage roll roofing is perfectly adequate. This is applied in the same manner, except that the overlap between strips is only a matter of a few inches. The total weight of the roof, obviously, is much lighter with this material.

Shingles, whether asphalt three-tabs, Dutch tabs, cedar shakes, fiberglass, vinyl, or aluminum, are generally used on pitches of four-in-one and up. Steel or aluminum sheeting can be used on six-in-one pitches and up.

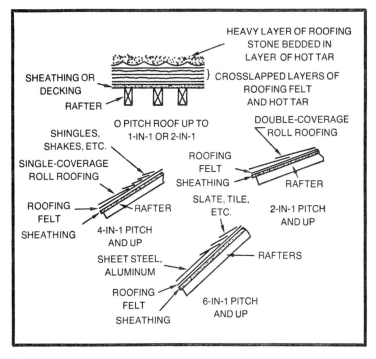

Fig. 5-18. The pitch of a roof determines the type of finish roofing materials that can be used; some kinds lose their effectiveness below certain minimum pitches.

The pitch of the roof determines both the cost of the roof materials and the amount of work involved in putting it together, too. A flat roof, for instance, has the smallest area for full enclosure of a given size of structure. Up to about three-in-one, there really is not much increase in the square footage, and then the roof area begins to rise rather rapidly with the pitch increase. A twelve-in-one pitch has nearly 1 1/2 times the area of a flat roof covering the same structure. The flat roof must be the most ruggedly built of all and so requires the most material. A steep roof requires less heavy framing, but more sheathing and weather surface material. Somewhere in between that good old happy medium turns up, as far as cost and amount of material is concerned.

On the other hand, a flat roof is the easiest to work on, little different than a floor. Up to four-in-one, the labor factor changes very little since at that pitch it is still possible to wander around in a relatively normal fashion and things don't have much of a tendency to go slithering to the ground. Beyond that point, though, the situation changes rapidly. Roof jacks have to be set, scaffolding rigged, safety harnesses and slings come into use. And a lot of muscle is required to haul everything into place, plus a lot of effort to put the pieces together while fighting gravity all the while. The very steep roofs, like those found on some Victorians, for instance, require a considerable amount of expertise and intestinal fortitude to assemble.

Besides cost and ease of assembly, there are other reasons for keeping the roof pitch down and making the total roof area as small as possible while still remaining in keeping with the rest of the structure.

In terms of initial cost, the roof is the most expensive section of the house shell on a square-foot basis. In addition, the roof usually carries the highest long-term maintenance factor, simply because it always takes it in the chops from the weather. So the less roof, the lower will be the maintenance costs and eventual replacement cost of the weather surface, and the lower the likelihood of leaks and damage.

Consider, too, that the roof is a great big radiator. The entire house is, for that matter, but the roof most of all. Any structural section angled away from the horizontal throws off heat from within at a great rate, primarily by conduction and by radiation. Don't forget that this works in both directions too. All that expensive heat that you generate inside the house during the cold weather is doing its best to escape into the atmosphere all the time. All that broiling sunlight that you don't need in the hot months is working mightily at getting inside the house. The smaller the roof, the less heat you will throw away in the winter, and the less you will absorb during the summer.

The roof, then, is probably the greatest problem of the whole house. At best, a long series of compromises is indicated, along with considerable pencil-chewing while trying to decide what is right. And on top of all that, the roof is architecturally the most important part of the house, frequently the most obvious and distinguishing feature.

While we are on the subject of roofs, there is one more item that needs a passing glance: the space underneath the roof. Like the basement room, the attic evolved as a matter of necessity. In the early days, before the advent of many of the roofing materials that we how have, the roof had to be steeply pitched in order to allow rain and snow to drain off easily and in order to present a tight and effective weather surface. The resulting space beneath the roof was unavoidable, and in many houses that space was considerable. Rather than letting it go to waste it was put to use, even though the space was neither as desirable nor as workable as the remainder of the living quarters. The process of trying to make the attic more livable spawned the use of dormers and such.

That situation hasn't changed much over the years; an attic is still substandard as living space because of the sloping lines and cramped quarters and the problem of easy access. When used for storage, an attic is inconvenient, becomes a catchall, can be a haven for bats and mice, and frequently presents a substantial safety hazard as well. About its only advantage is that you can handily see where the leaks in the roof are. And to top everything off, attics or under-roof crawl spaces are useless areas which cost you good money for no worthwhile return.

Unless the roof is flat, however, there must be some space between the rising line of the roof and the level line of the ceilings below. There are two ways to make the arrangement a bit more satisfactory.

First, keep the pitch of the roof as low as possible, commensurate with the type of weather surface you plan to use. The lower the pitch, the less space there will be underneath the roof line. Second, instead of building individual level ceilings in all the rooms, which is an unnecessary expense, leave the rooms open and finish the bottom of the rafters for your ceiling effect. This will open up the rooms and make them more spacious, aid in the acoustics department, eliminate the problems of attic heat build-up and moisture collection, cancel the need for attic ventilation, make the construction of the house a bit easier, eliminate a potential hazard, do away with access considerations, provide you with extra usable space, and probably cost you less money. And as far as any supposed loss of storage space is concerned, there isn't anything that can't be better and more conveniently stored elsewhere.

Doors, Windows, Walls, and Ceilings

6

In the last chapter I discussed the principal component sections of the house, the elements which form the bulk of the structure and make up the overall form. But there's more to a house than that; there are other factors of which you should be aware and which may have a bearing on your final design and on the cost and efficiency of your homestead

DOORS

Every house has to have doors, and some have a whole lot of them. Right at the outset, let me point out that doors, with their attendant hardware, are expensive, so if you want to save some money, here's a good place to do so. Not by buying the cheapest doors you can find, but by eliminating as many doors as you can from your plans to begin with.

As with almost everything else in the building field, you have a wide selection of doors from which to choose. Let's tackle the exterior type first. These are thicker than interior doors as a rule, the normal measurement being about 1 3/4 inches for exterior as against a 1 3/8 inches for the interior. They are stronger, and the extra thickness gives a little added protection from the elements, soundproofing, and insulation value. Most are made from wood. This can be solid or a solid frame with somewhat thinner panels inset in decorative fashion, or hollow-core where an interior honeycomb of light material is covered with a thin layer of wood veneer. The best type, however, uses a thick insulation core and a steel outer skin with a baked enamel finish. Glass panes, called lights, are available in various styles.

The lighter interior doors are available in many styles, including panel, full open louver, full closed louver, panel and louver, hollow-core, solid, and steel-cased. From a purely practical and safety standpoint, the steel door is the best again. It is fireproof, at least to a degree. The concern here is not so much that the door won't burn—it will, once it gets hot enough—but that it will retard the spread of fire far longer than a wooden door and allow extra time for escape.

Choosing door styles is as bad as picking out paint colors, and you'll have to struggle along as best you can with an armload of brochures from the lumber company. What is important to us right now is the size of the doors. They come in several standardized trade sizes. As to height, few homes use anything but the standard of 6 feet 8 inches, referred to simply as a six-eight door. Other heights are available if you need them, and sometimes the front entrance of a home is fitted out with a 7-foot or even an 8-foot door, or pair of doors.

The most used widths for passageway doors are 2 feet 8 inches and 3 feet , called a three-oh. Two feet 6 inches probably comes third, with the 2-foot size frequently used for closets.

Now measure your furniture. How many items do you have that won't go through the effective opening of a two-six or a two-eight? For complete ease of access, the three-oh door could be considered as a minimum size, and it will do no harm at all for you to use them everywhere in the house where there is any likehood of boosting bulky furnishings through a doorway. The cost of a three-oh is little more than that of a two-eight, but the convenience is considerably greater. And certainly the exterior doors should never be less than three-oh, with a half a foot wider deserving serious consideration.

Look at Fig. 6-1. This is the method almost invariably used in hanging doors today, inside or out. I've added the conventional storm/screen door as well. How well this door holds out the breezes depends on how carefully the door is fitted to the frame and how effective the weatherstripping is around the door. Here you actually build several frames; the double rough opening frame, the finish frame, the trim frame

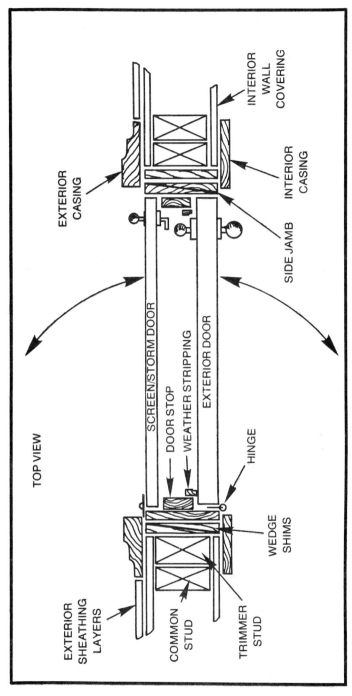

TOP VIEW

INTERIOR WALL COVERING

INTERIOR CASING

EXTERIOR CASING

SIDE JAMB

SCREEN/STORM DOOR

DOOR STOP

WEATHER STRIPPING

EXTERIOR DOOR

HINGE

WEDGE SHIMS

EXTERIOR SHEATHING LAYERS

COMMON STUD

TRIMMER STUD

Fig. 6-1. A typical exterior door installation as seen from the top.

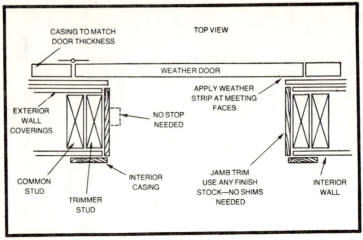

Fig. 6-2. The overlay door, originally called a batten door, is simple to build, uses a minimum of materials, and is relatively problem-free.

which acts as a door stop, and a weatherstripping frame. And any one of them can be out of whack. Furthermore, the frames will move around with heat and humidity and often with the weight of a constantly swinging, heavy door. The door itself will shrink and swell with the weather too. Much the same can be said of many types of screen/storm doors (the newer styles of prehung aluminum doors are quite weathertight, however), and on top of that you have to struggle through two doors, one probably with an automatic closer, with your sacks of groceries. That's as bad as the subway trains. And, of course, the conventional in-swinging exterior door takes up that valuable space inside the living quarters when left open.

Now look at Fig. 6-2. This type of door is called a *flush overlay*, which means that instead of being fitted to the narrow confines of a frame, it simply overlaps the surface and completely covers the opening with a substantial margin all around. You'll find this type of door on many kitchen cabinets, pieces of furniture, and all freezers and refrigerators. Simplest thing in the world to install, and there is no problem with fit. If there's some swelling or shrinkage, who cares? A rim of weatherstripping can be applied around the inner edge of the opening, or on the inside face of the door where it meets with the flat surface of the building, or both. You can install inside lock-latches which will pull top and bottom of the door up extra tight against the weatherstripping on stormy nights.

The door can be thick, a foot if you want it that way, without using up precious inside space. You can even build your own door out of heavy planking and decorate it with trim molding and studs or whatever, like a castle door. All kinds of possibilities. Hidden hinges and locksets (somewhat different than the normal ones) will make the place burglarproof, unless he's equipped with a chain saw.

WINDOWS

Windows are even more important than doors, and you'll probably have plenty of them in your house to concern yourself with.

A window serves one principal function: to admit light into the house. Back in the days of caves, the living was gloomy and dark. As man moved into living quarters of his own devising, that situation began to change as he cut holes here and there to let the light in. This, however, also admitted fresh air in copious quantities, precipitation to a lesser degree, and a smattering of bugs, bats, thrown stones, and errant arrows. In due course we discovered that by covering the window openings with a translucent material such as greased paper or animal tissue, light would come in but the weather wouldn't. But still, the light was a lot less, and such coverings interfered with a secondary window function which had also developed: seeing out.

With the advent of glass, the two-function window became complete. Now we had plenty of light and could still see out, but the weather and wildlife couldn't get in very handily. Of course, other folks could also see in to a certain degree, but that problem could be solved by putting the covers back into use, and those covers soon evolved into curtains.

In our ever-present quest for complete utilitarianism and labor-saving devices, we devised for the window a third and really unrelated function which until recent years has remained a convention. As houses became larger, more complicated, tighter, and more intricately built, we found it necessary to admit fresh air from time to time. So we made the windows openable.

Consider the labor-cost involved in making an openable window as opposed to a fixed one. First, you build a rough double frame in the structure of the house, into which the window unit will be placed. Then, someone at a factory somewhere makes another frame. Inside this goes a third frame, smaller and more complicated, which serves as finish trim, a stop for the window itself, and frequently also houses tracks or mechanisms for the opening and closing process. Then a fourth frame must be constructed, the sash to hold the glass. This too can be a rather complicated affair. Then the movable sash must be carefully fitted, properly suspended and attached to hinges and closure mechanisms, and provided with weatherstripping material which is at best of somewhat dubious value, depending upon both the type of window and the material, not to mention the quality control department of the manufacturer.

Now consider the fixed window. If a commercially made unit is to be installed, the rough framing still must be made in the structure of the house. Then, at the factory, an outer frame is fitted with the inner trim frame and the sash frame with the glass within. This is nailed up into a solid unit. At least a few steps have been eliminated, so there are some savings.

And last, look at the window made right on the construction site. In the simplest version, the wall studs or posts become the finished side jambs. Headers and sills of the same material are installed to form an open rectangle of whatever shape is necessary. Then if the opening is large or the design calls for several small lights (windowpanes), full-width dividers are placed within the principal frame to form a grid. A rim of small wood strips is then nailed around each opening to give the glass something to stand against. The panes are set, and another rim of strips is nailed around the inside to hold the panes secure, (Fig. 6-3). Simple, effective, functional, and inexpensive.

A large size, single-frame, partly openable window unit—that is, combining both fixed and movable sash after the fashion of many popular varieties of picture windows common to today's market—equipped with two layers of glass for insulation, could easily cost $10 a square foot, not installed. But note that smaller units where the entire sash is movable are far higher in price per square foot of glass area, something on the order of $25 a square foot. This figure would probably be close to the top end of the price range. But with the smaller units, you're not paying proportionately for glass area, but disproportionately for labor and hardware.

Commercial units with fixed glass cost only somewhat less than the movable type. A quick comparison of otherwise

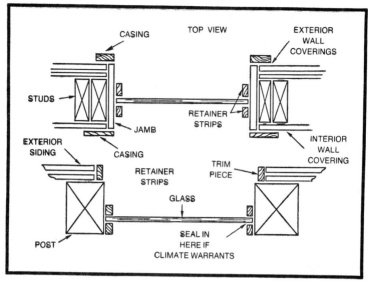

Fig. 6-3. Windows can be built quickly and cheaply on the site by simply trimming out the rough opening in the structure and adding divider strips, glass, and retainer strips.

identical units from a couple of companies shows a reduction of only about 20% in cost for large-sized units.

The window built on site is a different story altogether. The jambs are already there, a structural part of the house. The cost of sills, headers and dividers, plus the glass-holding strips is minimal even for a large window with a number of lights. Wood, after all, only runs about 25¢ to 50¢ a board foot, depending upon the kind. And glass is not terribly expensive, either, running from less than a dollar a foot for plain old window glass (which is perfectly adequate for most applications) to a couple of dollars for plate, and somewhat more for specialty glasses such as tempered or tinted. All in all, the cost for windows built on-site could easily be less than half the cost of commercial units. And since this type of construction easily lends itself to installation by the do-it-yourselfer, the out-of-pocket cash expenses could be a good deal less than that.

What about double glazing? The principal claim for double glazing—two panes of glass with an air space between them—is that this arrangement conserves heat. Glass has little thermal insulating value, radiates heat at a great rate, convects heat when the wind blows, and the total heat loss through the windows in a house is considerable. True statements all. Ergo, if you install double glazing, you'll conserve heat. Also true.

However, in some parts of the country the weather doesn't get cold enough to bother. And some people feel that double glazing, whether it saves heat or not, is of dubious value. This type of window is most expensive, and save for exceptional cases, if you work out the details carefully you will find that decades may pass before the additional cash outlay can be recovered in heat cost savings. If mortgage money is involved in the original installation, the economics are even worse. And this presupposes, too, that no drapes or curtains will ever cover the window area, which is in fact seldom the case in residential applications.

And there are some definite disadvantages. Double glazing helps to keep heat from leaving the house, but it also prevents heat from coming in. This means that you lose part of the advantage of the winter sunlight, so in the main what savings you gain at night you lose during the sunny days. In the summer, double glazing does keep some unwanted heat out, but that can be done much more effectively in other ways. Another point: every layer of glass before your eyes reduces incoming light by about 10%, and reduces visibility correspondingly. So you are paying extra for the privilege of reducing the prime functions of the window.

And another point: the air or partial vacuum between the panes may not remain stable. Impurities and moisture can form on the insides of both panes in cloudy patterns, and there is no way to clean them off. The only recourse is to remove the window and replace it.

But if you simply can't exist without double glazing, you can build your own on the site simply by adding another pane.

If you build your own on-site, single-glazed, nonopenable windows, you have saved a considerable amount of money and shed a number of useless, questionable, or bad features. But you are still left with three problems. First, keeping the cold out and warmth in (if that is a consideration where you live); second, providing ventilation for the house; and third, getting rid of the frost that accumulates from condensation in cold weather, or just the condensation itself in cool weather.

Taking the last first, condensation is admittedly much less of a problem with some of the more effective types of double glazing, especially the vacuum-filled kind (if the vacuum stays). Condensation forms only under severe conditions of very low outside temperatures and/or high inside humidities. With single glazing, though, condensation forms readily and of course turns to frost, sometimes in heavy layers, when the temperature drops below freezing. When this condensation runs off the glass and puddles along the window sill, two things happen. Unsightly stains appear on the wood trim if it is unfinished, or the paint flakes and peels. After a time, dry rot can set in, and the putty or other glazing medium can begin to disintegrate. In extreme cases, the panes can loosen in their beds, and the wind whistles through.

The obvious thing to do, then, is get rid of the condensation. Keeping it from forming is difficult in many climates though, and about the only way to reduce the problem is to maintain a low humidity within the structure. That too can be difficult. Another possibility is to deal with the condensation after it forms, simply by getting rid of it or getting it out of sight. If the bottom pane retaining strip is cut on a deep bevel, for instance, then the moisture runoff will accumulate in the pocket made between strip and glass, more or less out of sight, and will shortly dry up. This works best when the strips are made from unfinished or "natural" cedar or redwood, which is relatively unaffected by moisture. Any stains made will be out of sight. Another possibility is to drill drainage or "weep" holes through the bottom of the sill at strategic locations, which will allow the water to drain off rather than collect. These holes can be routed in such a fashion as to eliminate any consequential flow of cold air. In fact, one could go so far as to install drainage tubing down through the wall and thence to the outside. Or, or course, you can just forget the whole thing and live with frost on the windowpanes, as folks have done for generations.

Keeping cold out and warmth in is not a difficult chore. Some of this transfer can be due to direct leakage of air around the joints in the windows. This calls for a liberal applications of caulking at all joints, essential for factory-built window units that are installed in a rough-framed opening. Windows built on-site will have fewer such joints, or "crackage" as it is called, and those that do occur can be most effectively sealed up as the window is put together. Movable sash should be carefully and completely fitted with an effective weatherstrip.

The same is true of removable storm sash. The idea behind using storm sash is to provide a layer of dead air space between the two windows, and without well-nigh perfect weatherstripping and a complete seal, the resulting air movement between the windows will negate any advantages you might hope to gain.

The major amount of heat transfer, however, takes place by radiation from the glass, to whichever side is the cooler. The only way to combat this effectively is to block the radiation. The usual method is to use drapes or curtains. And since you probably will have these anyway, that should present no great problems. One thing you can do, however, is to make or buy your drapes with heat transfer in mind. Use those which have insulating properties, such as aluminized backing. If you prefer something more solid, and you live in an area of severe winter weather, you might consider a feature that our ancestors once used, called Indian shutters. These are heavy wood panels, fitted to cover the entire window opening on the inside. Some types are hinged to fold back in halves against the wall, others slide into hidden pockets within the walls, and they are most effective in reducing heat transfer. Folding louvered shutters, a popular decorative item these days, will also do a job, though to a somewhat lesser degree.

Now, about the ventilation. How do you get fresh air into the place, or cool the house down, if you can't open any of the windows? Not a difficult chore, really.

The first aspect of this problem is purely psychological. Most of us equate fresh air with a breeze. That's if the ambient temperature is above about 72°. Below that temperature, we call it a draft and cuss at it. But if you truly *must* be able to feel the wind in your face in order to be convinced that the process of ventilation and/or cooling is actually taking place, then you might not be happy with a house full of fixed glass, no matter what.

Once you have leaped this hurdle and are satisfied with fixed glass, there are several approaches to ventilation. If you live where the weather is hot enough to warrant air conditioning, then you are all set. When the air conditioning is operating, opening the windows is a no-no anyway. You can also use the outside doors, whose main function, unlike windows, is to open and close. Three or four wide-open doors, strategically located and even if screened, can provide plenty of air motion even in a large house.

Another method is to make use of natural air movement properties and ventilation ports placed in appropriate spots around the house. Every house has a cool side and a hot side. Cut a series of ports close to the floor on the cool side, and another series close to the ceiling on the warm side. With both sets open, the cool air will flow in at floor level, rise as it crosses the inside void, and exhaust as warm air through the high ports, creating a constant air flow and rapid air changes. When there is a breeze blowing, the effects will be even greater (Fig. 6-4).

The details of size and placement of the openings you will have to work out yourself, or have them engineered for you, depending upon the design of your house and the total amount of air you want to move in and out. About all I can tell you is that the exhaust ports should be approximately twice the size of the intakes, and you won't need nearly as much opening area as you might think. Building them is a snap. All you need is a hole in the wall, a permanent screen set in the middle of the hole, an outside, tight-fitting cover that you can remove in the spring and replace in the fall, and an openable inside door which matches the interior decor (Fig. 6-5). Each door can be insulated and weatherstripped if needed. Or you can do away with the outside door, move the permanent screen to the outside, and use one full-thick door openable from the inside.

You can also make use of a forced draft system, some of which may already be in the house. Vent fans in the bathrooms, range hood or vent in the kitchen, for instance. Turn these on and you'll get a substantial air flow. You can also place one or more fans in the attic or floor crawl space. Then by providing some open area for outside air to enter the

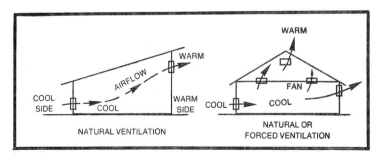

Fig. 6-4. Home ventilation can be accomplished by means of built-in vent ports, either with or without fans.

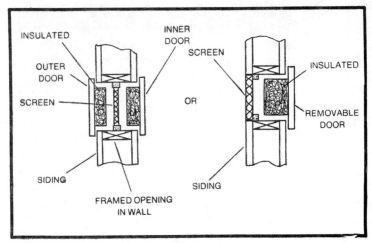

Fig. 6-5. Details of vent port construction. Doors can be hinged or clipped into place for easy removal.

house, you can move a tremendous amount of air. Those areas can be ports near the floor (they should be on the cool side of the house) or open doors, or whatever. To have variable air flow or volume capabilities, use variable or multiple-speed fans.

Well, how many windows should you have? Now, obviously you don't want to live in the dark. On the other hand, many modern houses have far more glass than necessary. Sometimes this can do more against you than for you. The alternative to glass is more wall. Wall is easier and cheaper to build, far more thermally efficient, stronger, and won't disintegrate when a golf ball strikes. You need just enough glass area to get the job done, but no more, because that would be time and money ill spent.

First, you have to have enough to fulfill any building code requirements that might apply in your area. This is usually expressed in terms of a ratio of so much glass to so many square feet of living space. Thus, if the rules say 10% of the square footage must be installed as glass, then a 2000-square-foot home would be required to have a minimum of 200 square feet of glass area.

You need to admit a satisfactory amount of daylight into the house, whatever that amount may be. There's no particular rule of thumb on this that I know about. Most people, however, will probably find that figure of 10% to be adequate. North side or heavily shaded windows have to be larger than those on the south or sunny sides for good natural lighting, so that percentage could well be increased somewhat here. And if the windows are double glazed, add on another 10% of the glass area (not the floor area). If they will be screened, add on another 30%. This will compensate for the amount of light lost by comparison with unrestricted single-glazed windows.

As to the second function of windows, seeing out, this is a matter of taking advantage of whatever views you choose to see. All this boils down to is that you should position the glass and make the expanse large enough to look at your views from various positions inside the house. The windows should not be so tiny or poorly positioned that you have to get out of your easy chair and crane your head through the opening to watch the rainbow. On the other hand, they need not be oversized or stretch from corner to corner of the room either. A common failing is to make up for the lack of planning by sticking in glass everywhere someone might chance to glance.

The important factor to consider is your sight lines. We all tend to look straight ahead, on the level, at least to begin with, and then adjust our sight lines from there. There is a further tendency for us to follow horizontal sight lines, rather than vertical. We scan, or to use a photographer's term, pan across one level plane, then jump to another, rather than seeing in vertical sweeps. And we usually start at the left and work towards the right, as we learned when we first began to read.

The angle of vision is usually rather slight too when we peer out a window, as opposed to the sharp angle we employ while watching a path before us as we walk. This is because the object or scene we are viewing through the window is seldom closer than 10 or 15 feet away, and most often a good deal farther than that.

This means, then, that a person of average height, standing and looking straight ahead, will employ a sight line about 60-68 inches above the floor, depending upon his specific height. And since most chair heights are all about the same, that sight line is on the order of 46-49 inches for a sitting person. For a full-vision window size, this gives us a top height of about 68 inches, and a sill level of about 46 inches. You don't always stand or sit right in front of the window to look out, however, so those dimensions should be increased somewhat to allow for both up and down angles of the sight lines as the viewer moves back into the room away from the glass opening. Seven feet makes a good, easy-to-work-with height for the top of the window, and still allows enough space for the upper casing trim, ceiling moldings, and drapery hardware in a room with an 8-foot ceiling. If the ceilings are lower, then the top of the glass should be lowered as well; 70—76 inches is still plenty of height, unless you do a lot of airplane spotting.

The sill height above the floor can be lowered to about 40 inches and give satisfactory results, and about 32 inches might be regarded as a low point. This will still allow you to place any average-size table underneath the window, since nearly all tables run about 29-30 inches in height. This also leaves plenty of room for coffee tables, magazine racks, baseboard heating units, ventilation ports, planters, or other relatively small household furnishings.

From the standpoint of normal sight lines, rectangular windows work better than square ones or tall skinny ones; we automatically find them easier to look out of, and they are more pleasing to the eye. Even if the sole function of the window is to admit light, such as a clerestory section, the vertical rectangle is a normal choice. One of the most popular styles of window, the double-hung sash, uses a pair or a series of vertical rectangles and does a satisfactory job.

The size of the individual panes is worth some thought too. If there are a great many panes in a small window opening, the overall effect can be one of confused busyness and restricted visibility. Huge panes, say 5 feet high and 8 feet wide, offer unrestricted visibility, but you pay a premium price for that advantage. Great expanses of glass do not give the impression of cozy, secure shelter or convey a feeling of contented security during nasty weather, as smaller panes and multiple mullions do. But on the other hand, there is no denying the airy spaciousness gained by great expanses of glass which virtually bring the outdoors indoors.

There is no ideal pane size, really, but those in the range of 2 by 3 feet to 2 1/2 by 4 feet seem to be just about right. These sizes can be arranged to form a large window unit in such a way that the divider strips are unobtrusive and do not interfere with the field of vision; you never even notice them unless you concentrate upon them. Such panes are easy to handle during construction, easily replaced if broken, need not be of great thickness or made of plate glass as large units should be, and both initial cost and replacement cost is low. They can be readily washed, even with brush and squeegee.

The windows in your house deserve a lot of thought and care in the designing. If you opt for factory-built units, you will automatically face some restrictions as to exact size, style, and design, because you must make your selection from the products offered. Arm yourself with a stack of catalogs from several different manufacturers, compare the various features, mull over the sizes and types, investigate the product quality, and settle upon those units that you feel will do the best job for you. And remember that the most expensive is not necessarily the best. If you build your windows on-site, then you will have to work out your own designs, but you will have more freedom to set the sizes and patterns exactly to your liking.

INTERIOR WALLS

We have already covered the exterior walls of the house, but most houses also have interior walls—a good many of them.

First, what are the primary purposes of interior walls? Principally, they provide privacy of two kinds: visual and audio. The first requirement is easy enough, since any opaque material, including a used horse blanket, will do the job. The second proposition is much more difficult, since sound travels all too readily. A third and lesser function of a wall is to demarcate various areas of the house—this is where I sleep, this is where you sleep, this is where we all eat, this is where the tools and the power saw will be used.

There are three kinds of house walls. The first is called a *partition wall*, or a nonload-bearing wall, and this is the most common type. It fulfills all three functions, though sound privacy is nearly always a great deal less than perfect. The second type is called a *load-bearing wall*, and fulfills the three principal functions as well as a fourth; it helps to hold the house up. At one or several points along the top of the wall, it supports roof rafters or second floor joists, or some other part of the structure. The third type of wall can take a number of forms, and is usually called a screen wall or a *divider wall*. This type serves to demarcate, provides only partial visual privacy, and virtually no privacy from sound.

This last type of wall seems almost always to be constructed to serve a particular need, complement a

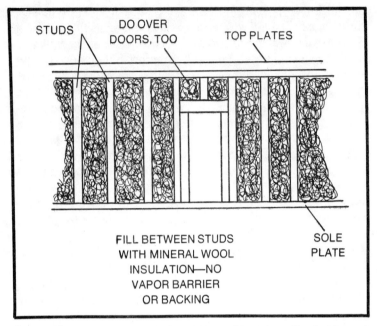

STUDS

DO OVER
DOORS, TOO

TOP PLATES

FILL BETWEEN STUDS
WITH MINERAL WOOL
INSULATION—NO
VAPOR BARRIER
OR BACKING

SOLE
PLATE

Fig. 6-6. You can achieve a modest degree of soundproofing just by stuffing typical stud wall sections full of mineral wool thermal insulation.

particular design, do a special job. The first two types, however, seem always to be constructed in the same conventional manner.

Interior partition walls and load-bearing walls are conventionally built in just the same fashion as the exterior walls. A series of 2 × 4 studs are parked upon a sole plate and secured at the top by top plates. Then a skin is applied, usually either plaster, drywall (Sheetrock or gypsum board), panelling of thin veneered plywood, composition board, or sometimes planking. Various bits of trim are then added to this, such as base molding, cove molding, door casings, and so forth.

The conventional partition wall does fulfill two of the functions nicely: it creates visual privacy and demarcates areas. But can it be made to fulfill the function of blocking sound as well? Yes, to a degree. This business of good sound blockage is a difficult proposition, and a compromise situation in nearly any house, unless a great deal of time and money and the hiring of a competent acoustics engineer are brought into the picture.

A conventional stud wall partition can be stuffed full of insulating material, as shown in Fig. 6-6. Sound will bounce off a hard surface, be absorbed by a soft surface. Sound will be transmitted by an elastic, rigid structure, but will not by a limp, nonelastic structure. So by stuffing the wall cavities—the material can be ordinary fiberglass insulation, sawdust, or steel wool—you will partially muffle some of the sound that is transmitted, and the noise level will drop. This works fairly well.

Another method is to use staggered 2 × 3 studs on 2 × 4 plates, as shown in Fig. 6-7. Then secure half-thick batts of insulation on each side. With this system, no part of the wall structure and skin on one side touches the other except at the extreme top and bottom. This lessens the sound transmission considerably, while the insulation muffles that which is transmitted by the skin. This is a reasonably effective method, one often used in houses.

A third system is quite successful, but also costly, space-consuming, and unlikely to be found in many houses. I've included it, though, in case you have a trombonist in the family you would like to sequester. Start with a pair of 2 × 4 or even 2 × 6 conventional stud walls separated by a space of a couple of inches, or more if you would like to use that otherwise empty space to run plumbing or ductwork through (Fig. 6-8). Make sure that nothing in the middle can touch either wall. Then stuff each wall cavity full of mineral wool

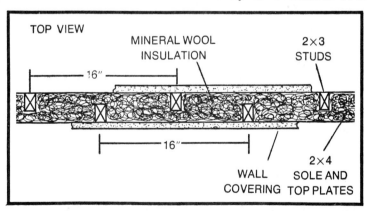

Fig. 6-7. Where sound transfer has to be minimized, use a stagger-stud wall like this one. The two wall skins touch only at extreme top and bottom, and the mineral wool stuffing serves to absorb most of the vibrations before they can pass through.

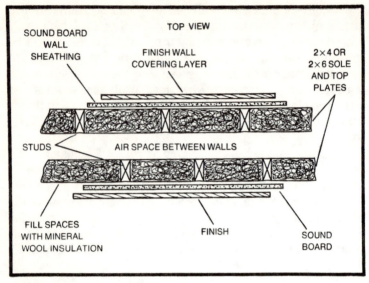

TOP VIEW

SOUND BOARD
WALL
SHEATHING

FINISH WALL
COVERING LAYER

2×4 OR
2×6 SOLE
AND TOP
PLATES

STUDS

AIR SPACE BETWEEN WALLS

FILL SPACES
WITH MINERAL
WOOL INSULATION

FINISH

SOUND
BOARD

Fig. 6-8. A full sound wall like this one will diminish any sound transfer to virtually zero.

insulation. Cover each outside surface with specially made sound board or acoustic board, which consists of fibrous material compressed into soft sheets. Then apply the finish wall skin over this. The result will be about as soundproof as you can get without doing a lot of highly specialized construction.

So far, all that I have said about partition walls also applies to load-bearing walls. There is one major difference between the two, however. With the load-bearing wall, the size and bulk and strength is more easily justified because it must be rugged enough to hold something up.

However, you can design the shell of your house to be self-supporting without need for any props. The structure built without supporting interior walls may well be superior in strength, as a unit. The costs may also be lower. Use of the interior space will be more flexible, not restricted by the presence of a support structure which can only be in one given place, underneath that which is to be supported. A partition wall can be removed should the need or the desire arise, albeit the job is a big one. A load-bearing partition, however, usually cannot be removed at all, short of tearing the house completely apart.

The best way is to let the roof hold itself up. If that simply is not possible, consider the use of support columns or piers which can be unobtrusively worked into the interior design, or perhaps a combination of an additional longitudinal supporting beam, in turn held up by posts or columns. Even this arrangement will give you more flexibility and less restriction of interior design than a load-bearing wall.

Assuming that you do choose to use the conventional stud interior walls in your house, there are a few things you can do to make the walls do more work for you.

Suppose that at certain prearranged points—and they have to be carefully planned—you frame openings into the wall structure as it is built, as in Fig. 6-9. One side of the opening is then covered in the usual manner with the wall skin. On the other side, apply the wall skin *around* the opening. Then trim the opening out with finish stock and moldings in whatever way suits you, and you have a built-in wall case. This sort of space is ideal for storing glassware, displaying collectibles,

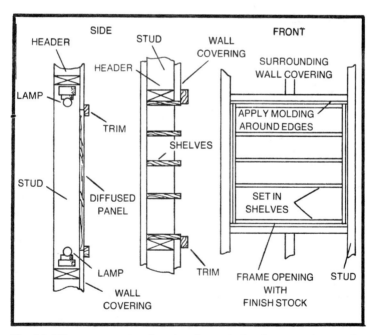

Fig. 6-9. Where conventional stud walls are built, oftentimes they can be put to use, even if of the load-bearing type. Shelving, shadow boxes, single- or double-sided light panels, or illuminated decorative panels, stained glass, or photographs are a few possibilities.

hanging rifles or pistols, or any such use. The opening can be built out a bit from the finish wall surface to provide a total depth of 6 inches or more, sufficient for a number of items, and solid or glassed doors, or sliding glass panels, can be added easily for dust-free storage. With light beamed down from a ceiling fixture, or even from a fixture recessed above the opening, such wall cases can be most attractive accent pieces.

Using the same sort of opening, you might cover one or both sides with a trim frame holding a piece of translucent plastic. With a couple of lighting fixtures properly placed within the opening and shielded to do away with shadows, you have introduced some soft indirect lighting to one or both rooms. This works quite nicely with rectangular shapes at midwall or above, or with narrow floor-to-ceiling units. The translucent panels can include a design for additional effect.

That 4 inches or more is also plenty to hold a built-in toaster, or an ironing board that swings down (or up), or a drop-down table or counter surface. You could line the interior of an otherwise useless wall section with narrow shelves, cover the opening with doors, and use the space to store canned and bottled goods. With a whole wall full, you'd have a skinny pantry, with room for a tremendous assortment of goodies, all stacked one deep and easy to get at.

There are plenty of possibilities along these lines; all you have to do is let your imagination roam for a while. Every time you comtemplate a partition wall on your plans, think: what can I possibly hide, or display, in all that space that would be wasted otherwise?

The third type of conventional wall, the divider or screen, is a most useful type which has been around for centuries. But the idea has limitless possibilities. Among the better known dividers, most of which are fixed but can be uprooted without too much trouble if need be, are such items as planters, bars, counter islands, half-walls, freestanding music or book walls, and decorative screens.

By using a divider, you can often do away with a permanent stud wall structure completely. Let's say, for instance, that your present plans call for a dining room and a living room separated, except for a wide doorway, by a conventional wall. This is rather stiff and formal, and you'd like something a bit less so, giving a somewhat more spacious aspect to a couple of rooms which would otherwise be just two more cubicles. Remove the wall entirely. In its place, build a box structure (Fig. 6-10) a little shorter than the original wall and only about 3 or 3 1/2 feet high, perhaps a foot thick. Match one side to the living room decor, install shelves or drawers or both in the dining room side, and use the interior space for storage of linens or crockery or whatever. Line the open top with a long copper or fiberglass tub, and stuff it solid full of house plants. This will give you a short wallscreen of varying height for a transitional but informal break between living and dining rooms. To make the screen more complete, rig some poles or narrow trellises from the planter to the ceiling and plant climbers. Or secure an anchor beam to the ceiling and suspend a number of hanging plants directly over the planter so that the greenery growing up mixes with that growing down.

There are any number of variations on this theme which can be put to good use all over the house. Another example: an open loft area overlooking a living room. The room is used

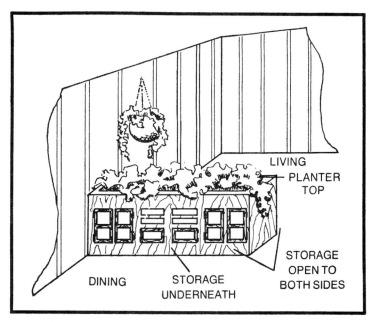

Fig. 6-10. Instead of bothering with conventional partition walls, a freestanding half-wall containing shelf or cupboard storage, drawer space, or planters can sometimes be used with great effectiveness.

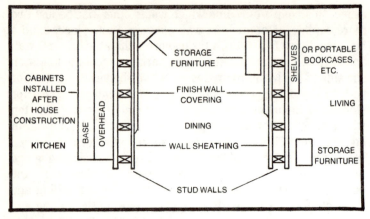

Fig. 6-11. In conventional construction, stud walls are built to divide the rooms, sheathing and decorative skins are applied, and then cabinets and built-ins are installed later.

most of the time for an activities area, but also as sleeping quarters for occasional guests. Rather than wall the loft off, in which case it would lose its charm entirely, install a series of traverse drapes which can be pulled across the opening as needed.

Check over all of your plans to see where you might be able to discard a conventional stud wall in favor of a divider or screen. If this notion appeals to you, you might be able to save some time and money and add to the character of your home as well.

There is another approach which I call substitute walls. Substitute walls, which are not really walls, fulfill all of the functions of the conventional wall and add a few more.

In Fig. 6-11 you'll see the conventional system of setting up interior partitions in a house. As the construction proceeds, stud walls are erected between the kitchen and the dining room, and between the dining room and the living room. Then, later on, a stack of cabinets is installed along the kitchen wall, with a finish skin applied on the dining room side. Most likely the dining-living room wall will be covered with a decorative skin on both sides, later to have furniture of various sorts shoved up against them. Why not eliminate those walls completely, and do something different?

First, the kitchen. Draw a line where that partition would have been, and start building cabinets. Attach a full row of wall cabinets, without any backs, to the ceiling joists or to supporting members nailed to the joists. Just let them hang there, built ruggedly and anchored solidly—they'll hold a lot of weight eventually. These cabinets can be constructed starting from the ceiling and working down, or as a complete unit put together on the floor and then boosted into place, or as a series of individual units, raised into place piecemeal and then connected to one another.

Then, directly underneath, build up a set of base cabinets, also without backs, in whatever fashion suits your plans. When everything is finished up and the countertop put on, nail sheets of plywood or particle board to the backs of both cabinet sets, full length, floor to ceiling. Or you can leave a section above the countertop open for a pass-through. Or you can cut some cabinet doors into the dining room side, making some of the kitchen cabinets accessible from both sides. The final step is to apply the finish, whatever it might be, to the dining room side of the panels. And there you have your kitchen cabinets and your dining room wall, all of a piece, and simply and easily constructed with no waste material, space, or effort.

For the living-dining room wall, the process is the same, though the function of the substitute wall differs. Here, instead of cabinets you might stack up open shelves, cupboards, hi-fi and TV enclosures, or whatever else strikes your fancy and in whatever combination seems appropriate. And rather than making the whole affair face into the living room, you could make some of the space accessible to the dining room for dishes and such. The final result would look something like Fig. 6-12. And note that all the functions of a wall are fulfilled. You have eliminated a wall structure and provided yourself with storage and display space that you would have to put somewhere anyway.

You can do the same thing with bedroom walls by dividing the rooms with closets, as in Fig. 6-13. Deep closets full of clothes, by the way are virtually soundproof. You can use a closet as an entryway divider. Full bookshelves would make a fine library wall, shelving and cupboards would work nicely for a rec room wall, storage closets and shelves for a mudroom wall. The possibilities are boundless. And incidentally, there is no need for the walls of closets themselves to be of conventional stud construction either.

A further extension of the substitute walls is to make them in modular sections, just a tad lower than the height of the

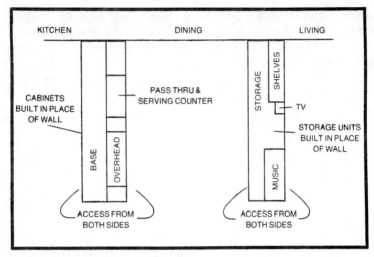

Fig. 6-12. Substitute walls eliminate the need for the stud walls, while fulfilling the normal wall functions perfectly well.

ceilings. You can even mount them on wheels. Either way, they can be shoved around here and there to provide a wall wherever you think you might need one at the moment. Your interior space becomes completely flexible—you can even move the bedrooms to the other end of the house if you feel like it.

CEILINGS

The ceiling is the forgotten part of the house, for the most part. A ceiling usually doesn't have much use, doesn't contribute much, doesn't get much attention except for a lot of cussing when it needs washing or repainting or the upstairs sink overflows. Mostly it's just a flat expanse of white or off-white with or without a bit of texture or trim.

There are three potential purposes of a ceiling. First, in many homes the ceiling serves as a finish skin which hides an unfinished and unattractive part of the house skeleton. The skin may be applied directly to joists or rafters, or to furring strips attached to joists or rafters, or may be dropped below the level of the skeletal members and attached to a subframework or a suspended metal gridwork. The second purpose is purely decorative. The ceiling may be made to blend in unobtrusively with the rest of the decor, in which case it goes virtually unnoticed most of the time. But in some few cases, the ceiling may serve as a decorative focal point, enhancing the decor of a room. The third function is actually made up of several miscellaneous items, depending upon the type of construction and the design of the house. A ceiling may serve to hold thermal insulation in place, or provide sound insulation, or improve acoustics, or provide a small amount of thermal insulation by virtue of its own composition. And undoubtedly it will also support a few lighting fixtures and perhaps loud speakers.

There are three conventional types: plaster, gypsum board, and tile. The first two are heavy, which means that the joists must be larger than otherwise would be necessary in order to support that extra weight. If there is a floor above,

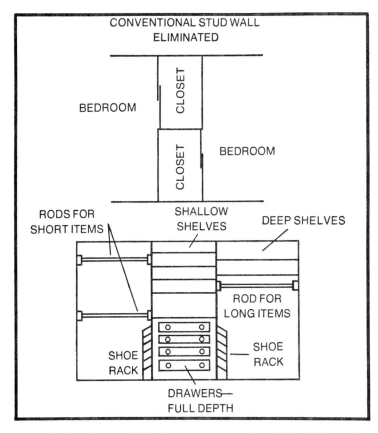

Fig. 6-13. Substitute walls between bedrooms do a better job than stud walls, while at the same time providing essential storage functions.

additional stiffness is necessary to prevent cracks from appearing in the ceiling surface. They probably will anyway, in due course, from unequal expansion and contraction of the different materials, shrinkage from aging, and normal vibration. Neither plaster nor the taped joints in Sheetrock can hold up perfectly to that sort of thing.

Plaster is probably the least desirable, with gypsum board next. Tile is the easiest to install, especially for the do-it-yourselfer, and the least expensive. There is now a wide variety of tile ceiling systems which can be used in any part of the house to good effect, and without many of the drawbacks of the other two ceilings. They are also easy to maintain and replace.

And you might also consider the alternative; no conventional ceilings. You can choose your structural materials so the interior surfaces, either as is or with the application of trim and finish, can serve as ceiling surfaces as well. For instance, in an open-plan house with cathedral ceilings, the rafter beams and the underside of the roof decking become the ceilings, with no need for any further covering. The same system can be used in individual small rooms, even in a two-story house. By simply building the structural materials up in the proper fashion, and then applying trim and finish as necessary, the need for building ceilings is voided.

7

Interior Design

Once you have made all your basic major decisions about your house, you'll want to consider the details of interior design. Perhaps the most important point to keep in mind is that the influence of interior design is pervasive, throughout the house and the entire homestead. The impact is tremendous, the rendition is entirely personal, or should be, and the time to consider at least a good share of these details is before you build, not after.

BASIC CONSIDERATIONS

Interior design depends, in part, on the way the house is constructed. First, there is the skin method. This is used with most conventionally built homes, whereby a decorative layer is attached to the skeleton of the house (Fig. 7-1). Once the framework is up, it is then covered with further layers of material, whose principal function is to hide the skeleton and give the finished, decorative effect. A cosmetic approach. The floor is built up with a subfloor, sheathing or underlayment, and finish in the form of carpet, tile, or finished hardwood, or perhaps masonry materials. The walls are plastered or covered with gypsum board, then touched up or taped, then sanded, then sealed or primed, then covered with layers of paint or wallpaper. Or perhaps thin wood panelling might be put up. Various moldings and trim pieces are added to hide joints and gaps. Assorted finishes are applied here and there.

The second method does away with the decorative skin, and instead makes use of whatever is there to begin with. The most obvious example that comes to mind is the log cabin, where the inside faces of the logs also serve as the interior wall surfaces, as is. If the ceiling is left out, then the log roof rafters

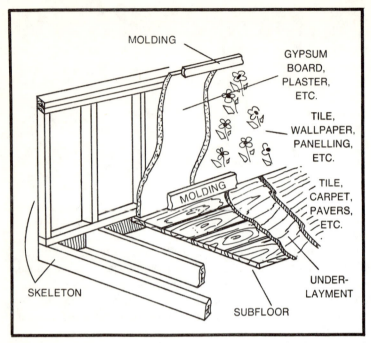

MOLDING

GYPSUM BOARD, PLASTER, ETC.

TILE, WALLPAPER, PANELLING, ETC.

TILE, CARPET, PAVERS, ETC.

MOLDING

UNDER-LAYMENT

SKELETON

SUBFLOOR

Fig. 7-1. The cosmetic, or applied, method of decorative finish consists of layer upon layer of materials, often nonstructural, each with something to hide and the last providing the decoration.

and their supporting trusses, together with the underside of the roof decking, serve the decorative function entirely with structural materials and without the addition of anything extra. Exposed post-and-beam construction also lends itself well to this type of decorating (Fig. 7-2). And though not all homes can be decorated in this way, still it is possible to stretch the decorating dollar by making use of both methods wherever possible. In a conventionally framed house, for instance, a one-layer floor could easily be used, thus doing away with the finish skin.

This means, then, that there are two approaches to the interior designing task. One is to build the house, plaster or Sheetrock the interior from top to bottom, and then figure out what to do in the way of trim, moldings, floor finishes, wall coverings, and ceiling treatment. The second approach is to figure out ahead of time just what will be done, incorporate those details right into the house design, and then continue from there.

This business of interior design has many complexities and amounts to far more than just sloshing on a coat of paint or slapping up some wallpaper. Part of the process is contained in the building of framework itself. The process continues as the dimensional aspects of sheathings and subsurfaces are considered. Oftentimes certain substructures must be devised as well. Then the final finish coverings have to be chosen and coordinated with one another, and various bits of trim and hardware brought into the picture. At the same time, texture, form, lighting, colors and furnishings must be interrelated, so that eventually a pleasing and functional overall design is wrought. Not an easy task. To show you what I mean, I'll cite a few examples.

The structural aspect of interior decorating appears when no decorative skin is to be applied. In a one-layer floor, the thickness, width, and kind of wood must be determined before the floor is laid so that it can be installed as the house is built. It is in place long before the remainder of the house is finished. No other layer of material will be placed on top of it, so the surface should be protected from stains and mechanical damage, a fact that the workmen must be made aware of.

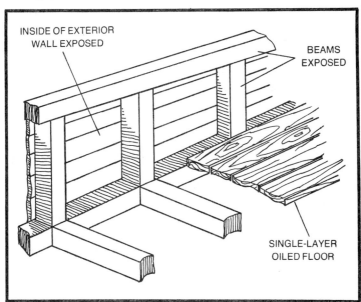

Fig. 7-2. In the integral method of decorative finish, the surfaces of the structural elements, as well as their form and texture, provide the decorative effect.

Perhaps a protective sheet of plastic or heavy paper should be temporarily laid down. Knowing what the final finish will be, such as floor oil, deck paint, or polyurethane, will make some difference in how much protection is necessary. And in addition, the joists and joist supports might well be engineered differently, and perhaps more economically, for a one-layer floor than for a conventional two- or three-layer floor. But you have to know all these details ahead of time.

To take another example, suppose you are using post-and-beam construction, and the wall posts will be exposed, as well as the interior side of the outside wall covering. This means that you must choose the materials used with this fact in mind. If you like the textured effect, then you will choose rough-sawn posts, and perhaps the wall planking will be the same. The posts might be of a dark wood, with the planks light. Or if you plan to paint the surfaces, they should perhaps all be smooth-planed. The size of the posts might be larger than structurally necessary, just to provide greater bulk and a feeling of strength and solidity.

Substructures come into play when an architectural feature having little or nothing to do with the structural integrity of the building must be installed as a part of the decorative scheme. Suppose that all ceilings in the house are set to 8 feet high, but in one room you would like a low ceiling. This might be to provide a warm, cozy feeling in a den or a library. This means that a subframework must be built to support a ceiling only 7 feet high. Or conversely, perhaps instead of having the 8-foot ceiling over all of the living room, you would like to open a part of the room clear to the roof line. This means boxing out the 8-foot high section, providing a high coverup partition to hide that section of attic or under-roof space you don't want seen, and allowing for some finish treatment for that part of the rafters and under-roofing which will remain visible.

Smaller substructures are often best built as construction moves along. These might include such items as windowseats, special stairwell treatments, framing for soffit or other special types of lighting, or some kinds of built-in furniture. In many cases it is difficult to tell just where substructures leave off and cabinetry begins, but in nearly all cases they are more easily planned and built right along with the house than they are at a later time.

The reason the dimensions of sheathing and subsurfaces have to be considered is simply because there is such a wide dimensional variety of finish surfaces which will be applied over them, and everything has to either come out even, or be precalculated so that differences will not be unsightly or haphazard but a part of the overall design. Flooring is a good example. Suppose the entire building floor is first covered with a layer of 1/2-inch plywood as a subfloor. You may have decided that some of the finish floors will be flagstone, some hardwood, some vinyl tile, and some wall-to-wall carpeting. If the flagstone requires an inch of depth, then you can lay that directly on the subfloor. If there is a great expanse of it, the floor joists may have to be made heavier, or additional supports put in. The hardwood, though, may be 3/4 or 5/16 inch thick. To make the two types of flooring even, you'll have to build the hardwood section up with another subsurface. The carpeting could be anywhere from 1/4 to 1 1/2 inch thick, depending upon its type and the pad used under it. The only way to save a lot of time and energy is to know well ahead of time just how you want to treat the floors.

Another good example is the walls. What combinations of subsurfaces and finish surfaces will you use in any given room? A conventional 2 × 4 stud wall with a layer of 1/2-inch Sheetrock on each side measures just about 4 1/2 inches thick. If you later decide to add a finish layer of panelling, you're immediately in trouble. The panels may be 3/16 or 1/4 inch, or if planking, 3/4 inch. If you already have the stock windows, however, the jambs only extend 4 1/2 inches, and you'll have a difficult job of trimming and finishing out ahead of you. Often it's just as easy to plan ahead for the things you don't know how you're going to handle as for the things you do.

Coordinating the final finish surfaces also deserves some planning. As mentioned before, with wall panelling you need to know the thickness of the panels. Solid planking used as wall covering is extra thick, and requires the installation of nailer strips to support it properly. If Sheetrock is used on the walls, later to be covered with wallpaper, all the taped joints must be reasonably smooth and flush. But if the Sheetrock will be painted, the entire surface must be well-nigh perfect if the finish is flat, and even better if the finish is glossy, because a glossy surface will quickly show up every tiny imperfection. On the other hand, if panelling is to be applied over a

Sheetrock wall, the joints need not be taped at all, which saves a lot of labor.

Trim is a general term used by most finish carpenters to denote any bits of extra woodwork which lend a finished effect. Moldings used around the edges of ceilings, mopboards, cove moldings along the floor lines, door and window casings, window sill edgings, chair rails around the walls, and valance boards are all in his catagory. Here the essential question in the planning stages is whether or not to use trim in any given spot, rather than the exact details of the trim itself. For instance, if you elect not to use any molding around the ceiling edges, that means that the drywaller must carefully fit the gypsum board where the wall meets the ceiling, and then the joint must be carefully taped so that the seam becomes invisible. There will be no molding to hide gaps or imperfections or misalignments. Another example: If you decide to use extra-wide door casings, say 6 inches or so, then doorways must be framed far enough away from adjacent walls or other openings so that there is room enough to fit the casings.

There are times when even the hardware needs some thought. Suppose you want a hardwood floor with no nails showing. You can glue the flooring down (which also needs investigation, because that doesn't always work), or you can toenail through the edges of the strips so that each row of nails is hidden by the next strip. But if you want to use visible old-fashioned nails, then you'll have to find a source of antique or simulated handwrought flooring nails and then set them carefully for the desired effect. Overlay exterior doors use unique hinges, and 3/8- or 3/4-inch overlay cabinet doors need a different type of hinge than flush doors. If you don't want to be able to see any part of the hinges, then both hinges and installation procedure are again different. Some hardware, such as doorknobs and latches, window locks and operating mechanisms, and many kinds of hinges have their own decorative impact and must be selected accordingly.

What all this boils down to is that a great many of your decisions involving house construction can have an effect upon interior design. The object of the exercise is to sort out as many of these details as possible ahead of time in order to avoid later patchwork, which is always less than satisfying. If you know exactly what you are going to do and what you need in the way of materials and hardware, you can keep a sharp eye peeled for sales and good deals and perhaps save a few coins. And if problems or delays in obtaining the necessary materials arise, you will still have some time to change plans or materials or to straighten out the problems without hindering the construction timetable of the house.

APPLIED FINISHES

Applied finishes, such as paints, stains, and varnishes are practically a way of life with us. We've been conditioned to them all our lives.

To begin with, different paints are used for different purposes, and a wrong application usually leads to later problems. So each individual finish should be matched to the job at hand; some kinds will do equally well in several different sorts of application. Paint, however, is not truly a wood preservative, regardless of what you might hear to the contrary. It is a cosmetic, used to create a decorative effect. If wood is subject to erosion by wind and weather, paint will indeed act as a mechanical "cover," and wear away instead of the wood. This works just so long as the paint cover is kept renewed, but mechanical protection isn't much of a consideration inside a house.

As a cosmetic, paint works well if properly applied and gives us ample and inexpensive opportunity to change colors often. It also covers up blemishes, knots, and wood fillers or putty so that less expensive woods can be used in many cases. But paint also requires maintenance, which varies with the specific type. And once a coat of paint is applied, there is no recourse except to continue applying more coats of paint as necessary. There is no turning back unless you want to undertake an arduous stripping job. Whether or not the finish should be flat, semigloss, or gloss depends upon where the paint will be used, and the preferences of the decorator. High-gloss enamel, for instance, cleans easily and is scrubbable, but also chips and scratches readily and reflects light annoyingly. A flat finish, or the other hand, does not reflect light and is easier on the eyes, but marks easily and doesn't take kindly to too much washing.

Stains work a little differently. There are two kinds, transparent and solid-body, but both consist primarily of oil with varying amounts of pigment thrown in. The transparent kind highlights the grain of the wood, while the solid-body type hides everything much like paint. Neither can be applied over a painted surface with much success, with a few possible exceptions in the solid-body types depending upon the specific case. Paint, however, can be applied over an old stain, so if you are not sure which you want, go with the stain first.

Though paint can be removed, stain cannot. It has much greater penetrating power, and once applied it can't be removed. Nor can stain be lightened. If you are not sure of the tone you want, always start as light as you can and darken as necessary. Solid-body stains will cover imperfections, puttied holes, and cracks in wood to a degree, but not completely, since for the most part it is lighter-bodied than paint. With a transparent stain, everything shows up, and defects in the wood may well be accentuated. This means that wood for use with these stains must be carefully chosen for the desired effect, nails must be hidden or countersunk and then filled with the proper type of filler, and the construction must be undertaken with great care. Knowledge of just how to apply and work with stains to achieve different results takes time and experience to acquire, but the results can be most gratifying.

Varnishes can be used over wood or over stained surfaces, since they are essentially clear. Again, any defects in the wood will quickly show up. Stain-varnishes eliminate the need for a separate staining process since the stain is already included, but the final effect is less controllable as to tone and shade. Used over bare wood or stained wood, varnish will provide a hard coating which resists spots and food stains and is readily washable and reasonably wear-resistant. If the ultimate result is to be transparent, then the wood base has to be carefully prepared and free from defects.

There are numerous other kinds of finishes. Plain boiled linseed oil is a good one, requiring little upkeep and a minimum of reapplication, while forgiving of stains, spots, scratches, and dents. There are also various specially compounded floor oils and stain-waxes made to be applied over properly prepared raw wood. The urethane finishes are plastic based, tough, resilient, colorless, glossy, and slippery. As with all of the applied finishes, the trick is to know ahead of time what you are going to use so that you can prepare for them. Exactly what you use will also affect the overall decorating scheme by way of light reflection, surface quality, acoustics, textural effects, maintenance, coherence of decor, and that indefinable quality often known as eye appeal.

FORM AND TEXTURE

Texture is the visual or tactile surface characteristics and appearance of something. Form, in this context, has two meanings. It is a visible and measurable unit defined by contours, such as a grouping of furniture; and it is the total effect of a finished room or a single item. Actually, there is no such thing as "correct" or "incorrect" form or texture or combination of the two; that is in the eye of the beholder. The importance of both to the designer at this stage is simply an awareness that both do exist and are forces which will come into play during the interior designing process, and that some of the decisions made now will, or may, affect both.

For example, highly glazed ceramic tiles present a glossy appearance with a smooth, hard texture. A wall done in these tiles has a rectilinear form consisting of relatively small identical geometric shapes. On the other hand, a wall done in small square mosaic tiles in a matte finish has a rougher texture, still hard, but by color arrangement can present a combination of rectilinear and curvilinear form in the same plane. The texture of a wall panelled with wood planking is entirely different from tile of any sort and varies with the specific kind of wood used and the finish applied to it, if any. The form is rectilinear, and may be horizontal or vertical or even diagonal, depending upon the lay of the planks and the grain characteristics of the wood if exposed. The addition of routed grooves in the planking, as with some of the standard decorative patterns like pickwick or barnboard, lends shadow effect and three-dimensional aspects, and may also add a curvilinear element to the form.

All of this becomes more than a little complicated when it comes to putting together a pleasing and compatible interior, and different combinations of forms and textures will impart entirely different visual effects. One room might be severely rectilinear in all of its shapes, with trim and moldings small and sparse. With the walls blank and smooth, woodwork painted in semigloss or gloss enamel, smooth and shiny hardwood flooring, ceiling lines flat and unbroken, indirect fluorescent lighting, and simple woodframe furniture with minimal upholstery, the room would be stark and simple, uncluttered and open, perhaps cold and cheerless, maybe even unfriendly and antiseptic. Much depends upon the feelings and the personality of the beholder.

On the other hand, a room whose basic lines are broken by nooks and crannies, moldings and trim which allow shadow effect and fuller dimensional depth, fitted with soft rugs or carpeting, hung with paintings or soft tapestries or draperies to break up open expanses of wall, equipped with scattered incandescent lighting, and furnished with fully upholstered pieces, becomes warm and cozy and has a softer and more comfortable feeling.

If you can get the basics of texture and form worked out ahead of time, you'll encounter fewer problems and disappointments later on.

COLOR AND LIGHT

Color and light play an extremely important part too, and they are tied together in terms of sensual impact. The colors you use will have different effects depending upon just how they are used. Visual and sensual impacts will also differ. A room done in dark colors will appear warmer than one done in light colors, but may also seem dismal rather than cozy. The lighter colors could seem cheerful rather than stark. Because of the varying energy levels of different colors, a given room can actually be made to look larger or more spacious by the use of light colors, or smaller and more compact by the use of dark colors. Either condition can be tempered by the use of form and texture; smooth light colored walls with vertical stripes give the impression of increased height, but dark and nubbly walls with horizontal patterns give a lowered effect.

Bright colors will increase the level of illumination possible from a given light source in a given room; deep and dark colors will do the opposite. Rooms with dark woodwork and furnishings might well be equipped with more glass area and more artificial illumination than one with light woodwork, ceilings, and furnishings.

Light reflection will play a part in illumination levels, shadow effects, and form, as well as spatial relationships. Numerous studies through the years have shown the floors should reflect 20–30% of the light reaching them. Walls should reflect about 30–60%, and ceilings should reflect about 60–80%.

Note too that colors do not remain stable to our eyes; everything depends upon the situation at any given moment. The paint chip that you choose under the fluorescent lights at the hardware store will not look the same in your home and will change from tone to tone depending upon where you place it. The rough-cut cedar planks you looked at in the noonday sun at the lumberyard will seem entirely different when you nail them up on the den wall, and change again under artificial light, one way for incandescent and another for fluorescent.

Little can be done about the effects of natural lighting, except for regulating the general illumination levels by placement of windows and by manipulation of drapes. Artificial lighting, however, is most important to the interior decorating scheme, and is a vital part of the home design for a number of reasons. Entirely different effects can be introduced by the use of different types and sizes of incandescent and fluorescent bulbs, and by using different methods of lighting. Further changes can be gained by installing a lighting system that can be controlled to produce variable combinations of lighting temperatures and levels for different moods and impacts. At the same time, the design has to be such that unwanted or undesirable effects are eliminated. For the best results , this should be done during the planning stages of the house.

ACOUSTICS

What you do in the way of both basic construction and interior decorating will determine what sort of acoustics your house has, and whether they are good or bad. And with the advent of whole-house piped music systems, this feature has become more important than ever before. And regardless of the canned music, good acoustics are essential to both physical and mental health. As is now being more widely recognized, the greater the noise pollution, the more problems people have in coping with their lives.

Sound bounces off hard surfaces. If the surfaces are hard, flat, and fairly good sized, sound bounces off them at a great rate. If the surface is elastic, as many walls are, more vibrations are set up and the sound passes on through. If the sound waves have a chance to bounce off this surface, then soon off another, and another, then it reverberates. The high frequencies, like a fingernail on the blackboard, are the most irritating. The low frequencies carry the best and the farthest and are the most readily heard and sensed. When the radio in the next room is playing a jazz tune, you can hear the string bass thumping even when you can't hear the clarinet or the trumpet.

Sound can be diminished by absorption and by dissipation. When sound waves strike a blanket, they are absorbed. The blanket is full of tiny crevices and air cells where the sound quickly loses itself. When sound strikes a heavily stippled ceiling, even though the surface is hard, it is so uneven that the sound waves rebound every whichway, much as light waves do. They scatter about, glance off at all sorts of odd angles from other surfaces, get all tangled up in one another, and finally subside. If the individual reflecting planes are large instead of small, much the same thing will happen provided that the angles between the surfaces are irregular and there are a lot of angular planes.

A cubic room with glossy panelled walls, a polished hardwood floor sans rugs, a plastered ceiling, and no drapes at the windows would be a nightmare of acoustic atrocities. A polygonal room with full carpeting, walls lined with loaded bookshelves, heavy tapestries and velvet drapes, a ceiling of stippled composition board, and filled with overstuffed furniture, would be somewhat akin to crawling into a cave stuffed full of cotton batting. Neither would be worth much as a music listening center.

If you are a music lover and consider listening to canned music in your home a serious business, then you have some further work ahead of you. In order to get the most from a fine stereo or quad sound reproduction system, the room where the listening will be done should be properly engineered to suit the system, and the system to suit the room; that's a two-way street. And for best results, that engineering should be planned, instead of becoming a patched-up afterthought which is less effective at greater expense. Consult a few of the many fine books available on the subject of proper sound system installation.

But acoustics are just as important to a home without a sound system. Fortunately this can be easily accomplished without any need for acoustical engineering, or the installation of special acoustical material, beyond those which have already been deemed necessary for specific sound-privacy situations. All you have to remember is absorption and dissipation.

The first capability you can easily gain by the judicious use of soft materials. Draperies, rugs, carpeting, upholstered furnishings, wall hangings and wool batt insulation within the walls will all contribute. Absorbent materials don't have to be everywhere; shouldn't be, in fact. Just in a reasonable ratio to the hard, flat surfaces.

The second capability is a matter of geometrics. Broad expanses of flat surfaces need to be broken up. Moldings and casements, especially the ornate ones, contribute nicely to sound dissipation, just as they do to light diffusion. Panels in the doors, shelves full of books and collections, paintings on the walls, nooks and crannies throughout the room area, plus lots of furnishings all help immeasurably. The more you break up the sound patterns and foil reverberation and resonance, the quieter the house will be.

THE SUPPORT SYSTEMS

Interior design also has to do with the support systems. These are the electrical system, the plumbing system, and the heating/air conditioning system. There is no need for you to engineer and lay out any of these systems unless you are interested in those fields or want to make the installations yourself. You can easily hire those jobs done; all are complex and complicated, and bounded by numerous rules and regulations. But each of these systems terminates in a series of usage points, and only you can determine what those usage points are to be and where they will be placed.

The usage points in the electrical system are called *outlets*. At every place where there will be some use made of electricity, there is an outlet of one sort or another. During the planning stages, make a note on your floor plan of all the places where you will want convenience outlets (plug sockets), where the appliances will be, and where the lighting fixtures will be. Spot every single piece of electrical apparatus and work out some of the details of what those items will be. You have a tremendous range of choices when it comes to user equipment in the way of lighting fixtures, appliances, receptacles, switches, and so on. At the same time, determine whether or not there will be any auxiliary systems like doorbells, security systems, sound reproduction wiring, extra telephones, intercom sets, and any special accessory equipment. Consultation with a book or two on residential electrical wiring might help you to get things properly lined up.

Much the same can be said for the plumbing system, and this is a little easier chore. All you have to do is determine what kinds of fixtures and accessories you want and where you want them. And if you can pin down the specific items by make and model number, so much the better. Here your concern will be with showers, tubs, water closets, lavatories and washbasins, kitchen and laundry sinks, water heaters, ice-makers, decorative fountains, sillcocks, washing machines, perhaps a lawn or garden sprinkling system, and any other items which involve the use of water. Once you have made the essential determinations about what you are going to have, somebody else can draw up complete plans for the system.

Heating and air conditioning pose a little different situation. If you plan to use a central or a split (part of the equipment located indoors and part outdoors) heating/air conditioning system, the first step is to make sure that you have allocated in your plans a place for the equipment. Then allowances must be made for piping, baseboard radiation sections, ductwork, registers, and grills. The overall design of the structure will dictate the placement of some of this equipment, but a certain amount of juggling can always be done. If you plan to heat with a series of unit heaters, you will have to determine where the units will be placed so they will fit in with your decorating scheme and at the same time do an effective job of heating.

Oftentimes many of these decisions can't be made until after the heating system is designed and engineered and you know more or less exactly what equipment will be used. Here again there is an incredible array of different equipment available. In this case, once you get the basics of the structure mapped out and know approximately what the house will consist of and how it will be put together, the next step is to design the complete heating system (or have it designed) and choose the equipment. Then you can relate this information to the sketches you have made up so far and make any changes or adjustments necessary so that the heating system will jibe with your interior design. Then when you make up the formal drawings, all of these details will be included and nothing will be left a-dangle.

Finished Drawings and Final Decisions

8

The last chore in the design process is to refine and scale your sketches into more or less finished drawings. You'll have to reduce and reassemble your notes into construction specification sheets. And from this collage of material, information final enough for an architect/contractor to use must emerge.

The finished drawings come first, and these will consist of a plot plan, floor plans for each floor or principal level, and a series of elevations showing all the sides or individual vertical planes of the structure. Unlike the previous sketches you have done, the finished drawings will be more formal, more detailed, and drawn to scale.

DRAWING TOOLS AND MATERIALS

At this point you will need a certain amount of equipment in order to do a creditable job, though a complete drafting setup is not necessary. The first requirement is a large, smooth, completely flat surface on which to work. It is possible to work at the dining room table, but this setup is decidedly awkward, uncomfortable, and inconvenient. If you anticipate doing a lot of drafting work over the years, you might want to invest in a professional drafting table, available from any drafting supply house. The obvious advantage is that there is nothing better to do drafting work on. Next best is a smaller portable drawing board on a stand, of the type that can be used for drafting, art work, or hobby activities. These are also available from drafting supply houses, or from some art and

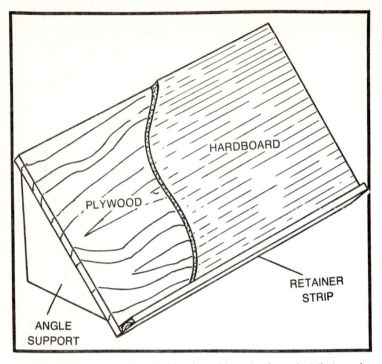

Fig. 8-1. Making your own drafting board is a simple project that can be put together from scraps.

office supply places, or for that matter, from Sears. About $50 will buy a nice one, adjustable as to angle and height, complete with built-in storage cabinet.

With either of these types of tables you will also need a tall stool about 30 inches high. Most draftsmen prefer a stool to a chair because they spend equal amounts of time standing and perched on the edge of the stool. Freedom of movement without constantly getting up and down from a chair is necessary, especially when working on large drawings.

At the bottom end of the scale, you can purchase a small drawing board—2 by 3 feet is a common size—which you can set and position anywhere you choose. They are inexpensive, and consist of just a flat board, often basswood, with a smooth, blemish-free surface.

Or you can build your own (Fig. 8-1). A chunk of plywood will do nicely for a base, thick enough so that it is completely rigid, and free of waves and warps. Nail the plywood to a couple of angular side supports so that the drawing surface is positioned at a comfortable working angle (about 30° to 45° from the level). Then secure a piece of 1/8- or 1/4-inch tempered Masonite to the plywood. You might also like to run a narrow piece of board across the bottom to form a lip to prevent things from sliding off the board.

Adding a large sheet of graph paper, which is ruled in both directions to form squares, makes the drawing chore much easier. There is a continuous grid of reference points always there to help you line things up and keep the angles and corners true. This is also available from drafting supply stores in a long-lasting vinyl-coated paper, which can be permanently taped to the drawing board.

The type of paper you will need is a special sort of tracing paper called *technical paper* for drawings and tracings. No other kind will do the job correctly. The drawings you do on this sort of paper are called tracings, which can be fed directly into a blueprint machine to produce the familiar blue and white copies. A large size is the easiest to use for house plans—17 by 22 inches is about the smallest that is workable, and larger sizes are better.

You'll also need a few tools to work with. Ordinary pencils will do the job, but drafting pencils work better. You can buy either the standard wood-cased type, which look much like a regular pencil without an eraser, or you can use a draftsman's lead holder. The lead holder is much handier to use, and there is a wide range of leads varying from extra-soft to extra-hard available to use with them. The hardness factor is a matter of preference tempered by the sort of job that is at hand, but for drawing house plans, number 2H or 3H works nicely. To keep the leads sharp, a constant process, you'll need a lead pointer. This can be either a small brass block with a renewable blade, or a larger and more expensive rotary type which mounts right on the edge of the drafting board. You'll also need a special nonabrasive type of eraser made for use with tracing papers.

For drawing the straight lines and angles, there are a number of possibilities. If you use graph paper beneath the tracing paper, then a couple of rulers will do the job. One should be long, at least 18 inches. The other should be an architect's scale. Such a scale has three sides and twelve scales, with all the necessary graduations needed to measure direct scale. The 1/8 scale, for instance, means that for every actual measurement of one foot you will use a one-eighth-inch increment on the rule, and the graduations are all marked out: 8 feet, 12 feet, 16 feet, and so on.

If you do not use graph paper beneath the tracing paper, then a T square, which you line up along the edge of the drawing board, will give you straight lines and proper right angles, provided that you have the edges of the paper lined up with the edges of the board, and that those angles are square.

In addition, one or several different sizes of clear plastic drafting triangles are quite helpful. So much so, in fact, that you'll probably find yourself using them instead of the rulers, any time actual measurements are not necessary. A plastic protractor will help you to determine angles, and if parts of your plans are curvilinear, a set of French curves in the form of plastic templates will save a lot of cussing. For regular constant-radius curves and arcs, all you'll need is a compass.

There are a great many plastic drafting templates available in all shapes, some of which are certainly handy but none of which are essential. There are templates for circles, squares, triangles, ellipses and such, in multiple sizes. There are others for electrical symbols, plumbing symbols, and standard architectural symbols. Amongst the handiest are those used for different furnishings, including kitchen cabinets and standard furniture. A quick look through the template racks at a drafting supply store will tell you what might be useful for your purposes.

With all the goodies gathered together, set up your drafting table where you can work under good light. The best light is strong and shadowless daylight. Failing that, light the drafting board with a combination of incandescent and fluorescent light. About the most comfortable arrangement is a 75-watt frosted or soft white incandescent bulb along with a 20- or 22-watt daylight fluorescent tube, positioned 3 feet or so above the board in such a way that no shadows fall on the work as you draw. There are special adjustable drafting lamps made in just this way. And you might as well get the lighting to your liking at the outset because this is close work that builds a lot of tension and strain and lasts a long time.

DRAWING THE FLOOR PLAN

To work. Position a piece of tracing paper on the board so its edges line up with the lines on the graph paper or the edges of the board. Precision is important, so that the finished results will not be cockeyed. Tape the paper down at each corner and at a couple of points along each side, if necessary, with tabs of masking tape. Incidentally, don't try to use ordinary painter's masking tape; it's too sticky and gummy. Use draftsman's masking tape.

The next step is to pick a scale for the drawing. The plot plan will use a small scale, since it must cover a large area. But the floor plan will use a larger scale. The larger the scale, the easier it is to use and the more detail and precision you can work into the drawing. If the house, or that part of it you are about to draw, is 36 feet long and 24 feet wide, and your paper is 40 by 30 inches, then you have room enough to use a scale of 1 inch to the foot. If the paper is only 17 by 22 inches, then you will have to drop down to 1/2 inch to the foot. Use the largest scale that will fit the paper, but you'll find it easier to pick one scale for all floor plans and stick to it. Use the same scale for all the elevations too, if you can.

The next step is easy: draw the first line. Doesn't matter much where you start, just follow your last floor plan sketch. Put in all the details, complete with door openings, windows, stairs, walls drawn to proper thickness; scale all the dimensions as you go along. There will be odd inches here and there that you won't be able to account for because the scale is not large enough to allow that kind of precision. That's nothing to worry about here because the scale is not followed in the actual construction. That problem is covered in the working or engineering drawings made up by architects or engineers. These drawings are done differently and call out specific dimensions in writing as necessary.

The purpose of the drawings you do is to make sure that all proportions and spatial relationships are correct and workable and that the layout is feasible. And though the degree of error involved here would be enough to raise hob with a carpenter's calculations, those few inches will make no difference to the present purpose. There are several "correct" ways of symbolizing the components of the structure; which should be used is a matter of preference or training.

When the form of the house is all blocked in, then you can start adding some of the details. One good place to start is with the kitchen cabinetry and the fixed or stationary appliances. Add in all the built-in units like bars, storage dividers, large planter units, shelving, and additional cabinetry in other rooms. Position the fireplaces, heating units, vanities, and plumbing fixtures, all to scale. By the time you have finished this process, you should have accounted for all the elements of the house that will be established during the construction and finishing process and will become essentially fixed parts of the dwelling. Then draw in all the major pieces of furniture that you will move in later. In addition, it is a good idea to include, perhaps with dotted lines, those major pieces which you plan to accumulate in the future.

Now comes the fun. Does everything fit? You may well find that you'll have to make some changes. Not only must there be room for all the various items, without any crowding, but there must also be space enough to insure complete usefulness of and access to all elements.

For instance, 3 feet between opposite counters in a kitchen does not leave sufficient work room. Four feet does. Is there room enough in the bathrooms to move around without getting trapped among the fixtures? The dining room table has to be set so there is a wide traffic area around it when several people are seated there. If the chairs are not to be left around the table, is there room to set them aside near the walls? The normal traffic patterns throughout the house should be clear and open, with no obstructions or awkward spots. You'll want sufficient space around the large furnishings in the living room to be able to move around them easily, and also to place end tables, coffee tables, floor or table lamps, and magazine racks.

Check all the heating units, heat registers, and cold air returns. There should be nothing blocking them, or near enough to them to cause airflow blockage or to create a fire hazard. Fireplaces have to situated away from combustibles and have enough clear space around them so that people can gather about to enjoy the crackling blaze. How about the sight lines from various points in the rooms? Centers of interest, like a fireplace, picture window, TV set, or conversation grouping of furnishings, should have clear and natural sight paths. In the bedrooms, none of the furnishings should be crowded. Easy access is important to easy cleanup and bed-making and clothing availability, as well as to wandering about in the dark.

Are all the entrances and exits obviously handy and easy to use? Stairways and landings convenient? Check all the door swings. Each should travel in the direction you want it to and have the hinges/knob on the proper sides. Make sure that when open, doors are in the clear, and when closed, access to them will not be obstructed.

Easy access to everything is most important to comfortable and friction-free living. All electrical panel boxes should be totally unobstructed. Openable windows have to be in the clear so you can open and close them without a struggle and get to them quickly when the rains suddenly start. Will you have to crawl over the TV set to get to the drapery pulls? Or stand on a ladder to reach the top shelf of the kitchen cabinets? Move the couch to water the plants? Crawl around on your hands and knees to fish out the good china? Dismantle the living room in order to fold out the sofa bed? Are there areas of wasted space that could be converted to some useful purpose?

Check through your daily living patterns, your activities patterns, your socializing patterns, your once-in-a-while living patterns like big parties or sick children or occasional two-week house guests. Will everything work as faultlessly as possible? If there are some things which bear further thought and perhaps some changes, either because they don't work or can be made to work better, now is the time to change them.

THE PLOT PLAN

Once the floor plan is all ironed out to your satisfaction, go on to the plot plan.

This too should be done in as large a scale as possible. Scale off the location of the main building relative to the property boundaries; orient the building to the right compass directions; draw in the outline of the building. Show the driveway and approach paths; draw in the outline of the garage, if it is not a part of the main building. Include any outbuildings or their major features such as a pool or a corral, in their proper relative positions.

Then draw in the septic tank and leach field or indicate the position of the main drainage line out to the street. Position the serving electrical power pole or connection point and indicate the incoming power line, whether overhead or underground, together with any extra poles that might be necessary. Show also any fuel storage tanks, whether above or below ground, or the gas supply pipe. Treat the water system the same way, showing the well location and the supply pipe or the water tap line out to the street main. The telephone line need not be shown unless it follows a different course than the power lines; usually the two are run together.

If there are any other major elements of your homestead that you would like to include, such as a vegetable garden or an

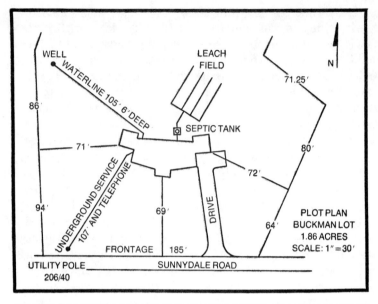

Fig. 8-2. A typical formal plot plan showing utilities placement and building setbacks. A larger plan would include more details and additional dimensioning.

irrigation ditch or stone walls or fencing, draw them in too. The more detail, the better. A *complete* plot plan will quickly show you whether or not you have a good layout. It will also show your banker, your contractor, your local building inspector, or your homeowners association, exactly what you are up to. In fact, presentation of this plan is a rigid requirement in a good many locales. A complete plan looks something like the one in Fig. 8-2.

THE ELEVATIONS

Now all that remains is a series of elevations, one for each face of the house. Following your floor plan and your notes, visualize what a wall of the house will look like, and have at it. Block in the outline of the wall, to scale, starting at ground level and working up past the foundation and to the eaves line. Add on the roof in perspective, if it is visible. Then set the doors and windows in their proper places and draw in the siding and roofing. If the pattern is complex or highly repetitive, you can simply do a small section and indicate in lettering that the finish siding will be cedar shakes laid X inches to the weather in random width and line (or whatever

174

the case may be), and the roof will be Jones #666 230-pound three-tab square-butt shingles, light tan, or whatever. Draw in any other trimwork or architectural features that you might have and call out in lettering any details you feel are pertinent, such as paint or stain colors. Figure 8-3 shows a sample elevation.

OTHER PLANS

You now have a set of finished design plans for your homestead, and in a great many cases, what you have is enough to go ahead with the building project. Many bankers will accept such a set as sufficient, and so will many building departments and architectural control committees and homeowners associations. A competent residential builder can also take the project from there. And of course, if you are going to do the building yourself, then you're all set. You have more knowledge about the details of the project than you can ever possibly put down on paper or impart to anyone else, simply because the whole affair is a product of your own efforts and imagination. At this stage, you've got all the material you need in the way of plans and specs.

But maybe your banker needs more plans, or you won't be around while the place is being put together by a contractor, or you want every picky little detail in black and white before you'll sign a contract to build. This will mean that other plans will have to be drawn. The exact number depends entirely on the complexity of the design. There may have to be a full drawing of the foundations, complete with dimensions and

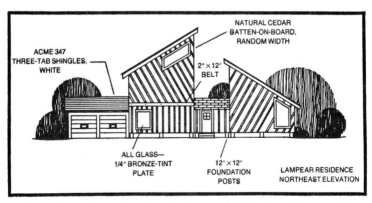

Fig. 8-3. A formal elevation showing the essential details of the overall appearance of the building.

details. There may be floor or roof framing layouts, and drawings of the house or parts of the house done in cross section, known simply as sections. Detailed drawings which focus on specific segments in large scale may be necessary. There may be separate sheets for the various subcontractors, complete in working detail to whatever extent is necessary. An electrical plan may be necessary, or a plumbing plan, or even heating and air conditioning plans. In complex projects there may be other trades involved to a degree requiring further plans, such as masonry details, roofing, or plastering. Extensive cabinetry or hand carving might require separate full sets of drawings with complete dimensioning and detailing.

So where do you get these plans? Your next move is predicated upon your own skills, expertise, knowledge, and interest. If you feel so inclined, you can go ahead yourself and make up whatever final working drawings are necessary. Otherwise, you'll have to turn over the material you have so far compiled to a competent architect or planner, and he'll take the project from there—with your guidance and for a fee. The net result will be a finished and professional job, but don't stop asking why. Examine all the work done, ask questions, and understand the conclusions reached. The information contained in these final drawings is what the contractor will be guided by, and what the drawings say is what you'll get when the house is completed. This is no time for surprises; know what you'll get in advance.

This is the time, too, for the final analysis. Go over every bit and piece of your dream home, now that it has taken its final form, and try to pick it apart again. Work out the explanations, make the changes, rethink some thoughts. When you no longer have any reservations about anything that you see on your plans, you are ready to build. Not before. But be aware of the fact that later on, as construction proceeds, you may well accumulate some further doubts or reservations.

MATERIALS LISTS

In order to determine what all the building materials are, specifically, and the quantities needed, you will need to work up some materials lists known as material *takeoffs*. This is not a particularly difficult chore, though it can be long and tedious. The trick is not to miss anything along the way. The procedure is simply to start with one section or element of the house and work your way through to the end, accounting for every item that will go into the construction process.

The framing is a good place to start. For instance, you will need so many linear feet of 2 × 6s for a foundation sill. You will need a certain number of 2 × 8s for floor joists. Count them up and note the necessary individual lengths. Then you will need so many sheets of plywood, whose characteristics you will have to specify, or so many board feet of decking or planking, which also has to be spelled out. Go through every part of the entire house this way, calling out in lists each bit of material and the specifications thereof. Include all the hardware, trimwork, fasteners, finishes, masonry components, built-in and fixed appliances, everything you can think of.

The electrical, plumbing and heating/air conditioning systems will carry their own material takeoff sheets. Unless you have designed the systems yourself, and are familiar with all of the component parts that go into the systems, request that the takeoffs and specification sheets be a part of the plans that are done up for you. Then check them all over to make sure that the usage items are the ones that you want and that their specifications are to your liking. These usage items include the lighting fixtures, wall switches, and convenience outlets and their finish covers, lavatories, sinks, tubs, showers, heat register grills or heating units—all the items that will be visible in the house. All of the countless hidden items which go into the makeup of the systems to render the usage equipment functional can be left up to the architect/engineer who devised the system, unless you have some special notions about some of the parts.

Oftentimes the easiest way to handle material takeoffs is to compile two sets of sheets. The first set consists of the materials needed for each *section* or *unit* of the building, in whatever fashion seems easiest or most obvious. You might have one set for forms and foundations, another for floors, a third for walls, another for roofing. Or if the building consists of several pods, or modules, you could proceed unit by unit. You might want to list the entire shell as a unit, and separate all the materials needed to complete the interior, from the studs out, on a room-by-room basis. Then you will know just what is needed for each section or phase of the project, and you can use those sheets as a guide to ordering and scheduling materials for immediate use. You will also know how the costs and material amounts are running as the construction proceeds.

The second set consists of an amalgam of the first, with no regard as to where the items are being used or when. To make up these sheets, just go through the first set and combine all the 8-foot 2 × 4s, all the sheets of gypsum board, all the rolls of 4-inch insulation, and so on.

The second task before you is to find out what all of these listed items will cost. A building supply dealer or lumberyard can help you here. You can simply give them a list of the materials, without any quantities mentioned, and request a unit price (per piece, per dozen or thousand, per square foot or board foot). You can then make the extensions yourself. When you get the prices, find out how long they will stand; prices gyrate so wildly these days that most dealers won't hold to a quote much beyond 30 days.

Going through all this trial and tribulation may seem like a waste of time, especially when you could probably get someone else to do the job. But it isn't. In the first place, this is your last chance to make sure that everything in your plans is the way *you* want it. In the second place, the process allows you to make sure that all the material, the quantities, and the specifications are correct, that the pricing is accurate and within the budget, and that the items are either readily available or can be obtained without long and disrupting delays. And in the third place, this is a good chance to make some changes, perhaps speed the construction process a little by making things easier, perhaps save a little cash or make the house a little better.

As you make up the takeoff and cost estimate sheets, do a little comparing, as to both specifications and costs. Perhaps another kind of construction-grade dimension stock than that specified is much more readily available from your supplier or is cheaper.

AFTER THE PLANNING

After planning comes building. But if you have never built a house before, be warned: the job is not easy. Countless hours are involved. There are many frustrations, especially if you work single-handedly. On the other hand, building your own place with your own two hands is a marvelous experience that brings with it a deep sense of accomplishment and satisfaction. If you honestly feel that you can do the job, dive in.

Failing that, you might want to do part of the work yourself and let others who are specialists, like electricians and plumbers, ply their trades at the same time. In this case, you will act as the builder, doing part of the job yourself and subcontracting the remainder. Or you can act as the general contractor, working as superintendant and coordinator, and hire most or all of the work done. This involves lining up the right men as they are needed for the various jobs: masons, concrete finishers, rough framers, roofers, plasterers or drywallers, plumbers, electricians, tin-knockers, heating men, glaziers, tile setters, carpet layers, the whole business. You're the boss.

Or you can pass the whole affair to your architect and have him put the job out to bid and then supervise the building process too. All you'll have to do is wait around for the keys to the front door and a big fat bill for services rendered.

Index

Made in United States
Orlando, FL
21 April 2025

60748103R00115